AF279151

The massive, eight-engined Bristol Type 167 Brabazon 1 (registration G-AGPW) flew over from its home at Filton, Bristol, for a flypast.

In the static park, 1949 (see page 29).

SBAC FARNBOROUGH

A HISTORY

PETER G. DANCEY

FONTHILL

Fonthill Media Language Policy

Fonthill Media publishes in the international English language market. One language edition is published worldwide. As there are minor differences in spelling and presentation, especially with regard to American English and British English, a policy is necessary to define which form of English to use. The Fonthill Policy is to use the form of English native to the author. Peter G. Dancey was born and educated in Bath and now lives in Milton Keynes; therefore British English has been adopted in this publication.

Fonthill Media Limited
Fonthill Media LLC
www.fonthillmedia.com
office@fonthillmedia.com

First published in the United Kingdom and the United States of America 2016

British Library Cataloguing in Publication Data:
A catalogue record for this book is available from the British Library

ISBN 978-1-78155-238-4

Typeset in 10pt on 13pt Sabon
Printed and bound by CPI Group (UK) Ltd, Croydon, CR0 4YY

Acknowledgements

The photographs used in this work are as accredited; unfortunately, there are a number where the origins are unclear. Illustrations and pictures are mostly from defunct partworks, *Flight* magazines and SBAC Farnborough Show programmes, and manufacturer 'hand-outs' or advertising literature including: Airbus Industrie GIE, Avro International Aerospace, BAC, British Aerospace, de Havilland, Eurojet Turbo GmBH, Handley Page, Martin-Baker Aircraft Co. Ltd, Rolls-Royce, Russian Armament State Corporation Rosvoorouzhenie, Short Brothers & Harland Ltd, Sud Aviation, and Westland Helicopters. To all contributors, I express my gratitude, in particular to those I have been unable to contact. To anyone I have forgotten, I offer my unreserved apologies.

Peter G. Dancey

Above: The Blackburn Beverley at the 1952 show.

Below: Crowds view the static display at the twenty-first show, 1960.

Contents

Introduction

Farnborough's association with aviation goes back almost 115 years, when the first balloon flights by Yeomanry Army Officer Capt. J. L. B. Templar, who persuaded the British Army in the military use of balloons, conducted his initial evaluation flights.

Trials conducted by Lt Glover of the Royal Engineers lead to a factory being set up in nearby Aldershot in 1890. This was later to move to South Farnborough as Her Majesty's Balloon Factory and Balloon Section Royal Engineers in 1905. A large balloon-erecting house was built in 1905 and the various Balloon Sections were amalgamated to become a Company.

Also in 1905, a purpose-built Balloon School was erected, which almost immediately, with the formation of the Balloon Company, became HQ Balloon Battalion in 1911. Whilst balloons were becoming established at Farnborough, Samuel Franklin Cody, from the nearby village of Ash, was undertaking man-lifting flight trials from nearby Laffan's Plain on behalf of the British Army. Cody was at the same time building an aeroplane in the Beta shed within the Balloon Factory. When complete, Cody undertook what was the first powered flight in Britain on 16 October 1908, flying a distance of 424 metres (1,390 feet). The flight ended in a crash; the port wing-tip of his aeroplane struck the ground whilst he was manoeuvring to miss a tree. Although Cody survived the crash, he was sadly killed on 7 August 1913 when his Waterplane crashed on the north-western corner of the present airfield, following probable structure failure of his machine in flight.

Nonetheless, it was largely due to Cody's pioneering flights and experiments that the military established the Army Aircraft Factory on 1 April 1911, with the Army Aircraft Factory becoming the Royal Aircraft Factory (RAF).

The Royal Aeronautical Society had been founded in 1866 and the Aero Club of the United Kingdom had been formed in 1901, two years before the Wright brothers' historic flight at Kitty Hawk, North Carolina. Aviation industrialists in Britain were already recognising the potential of the new heavier-than-air machines and attracted by the offer of free floor space at London's 1909 Olympia Aero and

Motor-Boat Show. This event provided them with the first real showcase for their risky, flimsy, and somewhat life-threatening machines.

The expense and perils of taking to the air at this time were well illustrated by the reports of a US journalist, Augusta Post, who wrote:

> ...the cost of new machines can range from $5,000 to $7,500, and the average cost of breakages while learning to fly can amount to almost $2,000, a not-insignificant amount, the pioneering aviator needing to be an extremely rich man.

Only one of the eleven aeroplanes on show at Olympia in 1909 (a French Voisin) had actually flown before, and it was perhaps just as well as this first British aerospace trade show lacked the facilities to demonstrate whether or not the products on display could actually fly. Aerial displays were, however, organised for later in the year at Blackpool and Doncaster, in the north of England. From the handful of pioneering aircraft builders present at these events, only one company—Short Brothers, which began manufacturing Wright biplanes in February 1909—still survives in the modern day, albeit now under the Bombardier banner. Only one other came close—Handley Page Ltd, which displayed a Hungarian-designed aeroplane at Olympia in 1909 and survived independently until 1972.

The customers for these early flying machines were wealthy young males who had faith in their own immortality. In reality, it appeared at this time that the new flying machine was nothing more than a rich man's toy. However, it was not long before the prospect of a global war changed perspectives, and with the military already evaluating and operating the new machine as a weapon of war (albeit in limited numbers), aviation soon went from strength to strength—literally, by way of the type and capabilities of the machines produced.

Large-scale military funding was suddenly available from governments, firstly to develop purpose-built reconnaissance types, then fighting scouts, and ultimately bombers. Aeroplanes were soon produced in their thousands.

In April 1912, the Royal Aircraft Factory at Farnborough was requested to concentrate solely on aeroplane production as opposed to balloons. A number of experimental types were produced, such as the Se.1, Fe.1, Be.1/Be.2, Fe.2 series, Se.4, and Se.5. Some of these were later produced by established aircraft manufacturers such as Airspeed, Avro, de Havilland, and Sopwith, especially when the Royal Aircraft Factory was ordered to cease aircraft manufacturing to concentrate solely on research, design, and development.

In total, some twenty-eight different variants were designed at Farnborough, which resulted in 533 aircraft being built on site. The rapid increase in aircraft production at the Royal Aircraft Factory demanded an increase to the number of skilled engineers employed there. A purpose-built housing estate for their wives and families was constructed at nearby Cove. As the number of aircraft produced increased and the necessary modification and development work was undertaking, increasing numbers of types began to be built by sub-contractors. Those pioneers who had ventured into the aircraft industry soon became millionaires. They went on to form their own trade associations, mainly to give the new industry a semblance of respectability.

The SBAC (initially named 'The Society of British Aircraft Constructors') was formed in April 1916, in the middle of the First World War. At this time, the organisation only had seven members, all of whom were private individuals. These were great times for the SBAC members—for example, over 18,000 Sopwiths were built over the course of the war. However, two years later, the November 1918 armistice brought a total collapse in demand, with thousands of war surplus airframes on the civil market. Nevertheless, the industry survived this rapid downturn in its fortunes and learned to adapt to 'peacetime' conditions. Most turned to producing the first civil passenger transports, seeking to replace the early converted military planes (mostly modified former Airco, de Havilland, and Handley Page bombers).

In 1920, the SBAC collaborated with British motor manufacturers in staging a second combined show at Olympia. In the same year, the first Royal Air Force Pageant and Air Display was held at RAF Hendon, north London. This became an annual event that was essentially held to show off military types, but it also included a static exhibition of some of the latest prototypes in what was known as the 'New Types Park'. This area always proved to be the focus of attention, particularly for overseas visitors, and the SBAC soon realised that it represented an excellent method of promoting the products of the aircraft industry in Britain.

From 1932, this section formed the basis for the SBAC's own small show, which followed the RAF Pageant. At the first SBAC-organised event, some thirty-four new aircraft were displayed. The show lasted only one day and entrance was limited to invited foreign guests. The event set the pattern for the future; in the morning there was an examination of the static aeroplanes, followed by a flying display after lunch. The show became an annual event.

The original airfield site at Farnborough, excluding the factory, consisted of approximately 275 metres × 370 metres (300 yards × 400 yards) of grass, from which the aircraft produced were flown. As aircraft became larger and heavier, there was a need for expansion; this gradually occurred in a westerly direction, with concrete runways being laid in 1942 and additional extensions in 1944 and the 1950s.

A further runway was laid later, intended for use by aircraft of the Naval Air Department, who sited two Larynx Mk 1 and Mk 2 catapults on it for carrier-landing trials. However, this runway became redundant with the opening of runway 18/36 in the early 1940s. Flexible deck-landing trials involving a DH Vampire jet fighter (using a large rubber mat) were also conducted on an old adjacent runway from 1946 to 1948. A modified flexible deck was later laid at Farnborough for further trials, and from April 1952 to June 1955, many successful experiments involving the DH Vampire and Hawker Sea Hawk were carried out.

At Farnborough, the Royal Aircraft Establishment (RAE) really came into its own with regard to aircraft systems, weapons, and other scientific research during and immediately after the Second World War. At the peak of its activities, the airfield could boast nearly 200 or so resident aircraft types, all undergoing various research and experimental equipment and development trials. These included experiments associated with general handling trials, high-speed flight, altitude research, fighter and bomber weapons trials, assault gliders, radio and radar equipment development, jet engine development, aircraft catapult launch trials, flexible deck landing trials, arrester gear, and angled deck research work. The latter investigations were closely related to the introduction of jet aircraft to carrier operations. There followed research regarding missiles, guided weapons, and spacecraft developments. It would be true to say that most of the current military and civil aircraft systems, such as communications, navigation and weapons, owe much to the experimental and development work undertaken at Farnborough.

The jet era saw work on turbine-engined aircraft forge ahead under the auspices of the Farnborough RAE Engine Department. The department was established in 1941 at a remote site at Pyestock. It was extended in 1946, adding the National Gas Turbine Establishment (NGTE), along with further site redevelopment that included a test-bed installation in the 1950s. Considerable engine research and development work was conducted here, while the early jet aircraft flew extensive trials and experimental work from the nearby Farnborough Airfield.

In the early 1960s, a new test facility known as 'Cell 4' was built at the NGTE to house the Olympus engine power plant for use in the BAC/Aerospatiale Concorde SST. This test cell was used exclusively for the Olympus engine trials and was fully capable of simulating supersonic flight test conditions, which were vital in the development of Concorde for examination of its flight and performance characteristics. Another interesting project to emerge from NGTE during the late 1950s and early 1960s was a proposal for a vertical take-off blown circulation controlled rotor passenger airliner. Although it was never built, this unusually configured concept led to further developments in this field.

Pioneering aviators, aviation engineers, and scientists had a very steep learning curve in the 1930s and 1940s. Many accidents occurred as a consequence, and the RAE found it necessary to conduct its own in-house accident investigations in relation to aircraft structures, materials, and metallurgical fields. An early pioneering investigation was related to the fatal accidents involving the de Havilland Comet 1 jetliner. Much work was also undertaken using the water tank related to the fatigue-proofing of the Bristol Britannia turboprop airliner, with an entire aircraft placed in the water pressure tanks.

Much of the early investigative work started during the Second World War continues today, and Farnborough has regularly assisted the present-day Air Accidents Investigation Branch (AAIB) in numerous areas of research and analysis. The AAIB, which is responsible to the Ministry of Transport, undertakes investigations into many accidents worldwide that involve British-registered aircraft or British-manufactured aircraft, both military and civil.

From 1939 onwards, the wreckage of shot-down enemy aircraft in the UK was taken to Farnborough for evaluation. Of particular interest were the engine fuel injection systems of many German aircraft. Gradually, increasing amounts of German aircraft wreckage were directed to Farnborough for investigation by the Air Intelligence Branch of the Air Ministry in London, who, together with the RAE and its 'parent', the Ministry of Aircraft Production (MAP), would then decide if the wreckage or complete aircraft warranted further analysis or investigation. This was to ultimately lead to the formation of the Enemy Aircraft Department, through which the Experimental Flying Department undertook extensive flying trials on these aircraft, often revealing many secrets of German technology to the British aircraft industry. One particular event of note involved a push-pull single-seat heavy fighter, the Dornier Do 335 Pfeil, on 18 January 1946. The pilot lost control of this aircraft when a rear-engine fire burnt through the control surfaces, and it crashed on a bicycle shed at nearby Cove School. The pilot, Grp Capt. F. Hards DSO, Commanding Officer of the Experimental Flying Department, was killed, but miraculously there were no casualties on the ground.

The flying of captured German aircraft was greatly curtailed as a result of this accident, and the majority were scrapped or taken away for disposal by the Ministry of Supply. Even so, it was not until 1955 that the last captured German aircraft, a Fieseler Fi 156 Storch STOL light three-seat observation aircraft, was relinquished by the RAE, the aircraft having had a useful life at Farnborough as a trials and communications aircraft.

Farnborough Airfield was the focus of much attention by Luftwaffe bombers during the Second World War. In an incident on 16 August 1940, eight Junkers Ju 88s attacked the airfield from a height of just 910 metres (3,000 feet), dropping twenty high-explosive bombs. Half of them fell on the RAE site and some fell on the town, killing three Home Guard personnel on duty—the first to be killed on active service. As a result of the attack, a Duty Defence Flight was formed consisting of a Gloster Gladiator and Hawker Hurricane and Supermarine Spitfire fighters.

As a result of damage to many of the RAE offices and workshops, many were subsequently dispersed to around twenty or so remote sites in the local area, where they remained for many years. In 1946, the Empire Test Pilots School (now at Boscombe Down, Wiltshire) moved from Cranfield Aerodrome, Bedfordshire, to occupy a site on the southern side of the airfield. Its residency added a further thirty or so aircraft to an already busy airfield. Amongst the types flown by the ETPS were DH Chipmunks, Devons, Vickers Viscounts, Hawker Hunters, English Electric Canberras, Westland Scout helicopters, Whirlwinds, Short SB.5s, and English Electric Lightning to name but a few.

Another aeronautical unit to move to Farnborough in 1939 was the RAF Physiological Unit, which had previously been based at RAF Hendon, north London. At Farnborough, the unit was initially housed in a timber hut situated alongside the old balloon shed near the Pinehurst cottages, but in 1945 it was moved to the site behind the present-day SBAC exhibition area, becoming the RAF Institute of Aviation Medicine (RAF IAM), as it remains to this day. This unit is still located at Farnborough, although it has been absorbed into the Defence Research Agency (now QuinteQ) Centre for Human Sciences.

The RAF IAM has operated its own aircraft throughout the years, undertaking research into aircrew and human factors. The flying element of the IAM relocated to Boscombe Down with its dual-seat Hawker Hunter T.7, which was the last Hunter aircraft in UK military service at the time. A BAe Hawk T.1 trainer was subsequently used.

Of particular interest was the anti-g tests carried out by the IAM, who used the Supermarine Spitfire operated in the UK for these experiments until 1955. Another area that the IAM specialises in is the 'desensitising' of aircrew who still suffer from air sickness after their training. This work is accomplished with the use of a Hawk T.1 trainer aircraft.

The IAM operates a man-carrying centrifuge at the Farnborough facility. It was commissioned in 1954 and is used for various aircrew high-g research tests. It previously had a maximum-'g' capability of 30g, but this has since been reduced to the 9g experienced in modern combat aircraft (such as the LM F-16 Fighting Falcon and Eurofighter Typhoon. It is believed that a new centrifuge for the training of Eurofighter Typhoon pilots is to be installed.

Farnborough has a long-standing association with the Meteorological Research Flight (MRF), which works in conjunction with the UK Meteorological Unit headquartered at Bracknell, Berkshire. Early meteorological flights were undertaken by the RAF, and during 1946 they used former wartime HP Halifaxes, DH Mosquitos, and later converted HP Hastings transport aircraft. However, the Met Office was to form its own flight using a number of EE Canberra light-bombers followed by a specially-modified Vickers Varsity that culminated in the modified Lockheed C-130K Hercules W.2 (serial XV208). The Hercules (which has since retired) was fitted with a 7-foot-long nose boom for scientific experiments, a weather radar in a special enclosure above the cockpit, and various other sampling devices. The aircraft, *Snoopy*, was delivered to Farnborough in 1974; it was a familiar sight there until it relocated to Boscombe Down on 24 March 1994. *Snoopy* was withdrawn from use in 2001, being replaced by a civil-sponsored BAe 146-300 Regional Jet that was actually the first -300 variant to be flown.

In the new millennium, Farnborough is a far quieter place than it was in the heyday of the British Aviation industry. Nevertheless, in 2008, sixty years after the SBAC made the airfield its permanent venue for the now biennial shows, millions of visitors still attended to view the latest heavier-than-air machines both on the ground and in the air, in addition to all the latest technical innovations from around the world in the exhibition halls and stands.

1

SBAC Hendon 1932–1935

The SBAC's Farnborough International Exhibition and Flying Display is recognised as the most complete exhibition of the year. For those who wish to buy or sell aeroplanes or products of the aerospace industry and for those who wish to keep abreast of the latest technology, Farnborough is second to none—'The Greatest Show on Earth'. No exhibitions were held from 1938 to 1945.

In 1948, the Society of British Aircraft Constructors (SBAC) moved its now-biennial show to Farnborough, Hampshire. This became its permanent venue; as the size of the show and exhibitions grew, it was allocated a permanent site on the south-eastern part of the airfield, on rising ground. It was very close to the spot where Samuel Cody had made his first flights. The SBAC is now known as the Society of British Aerospace Companies.

It was in 1932 that the first SBAC display took place at Hendon, north London. As an experiment, an extra day was added on to the annual RAF Pageant held there. Admission was by invitation only. Rather bizarrely, the society instigated a ban on photographing the exhibits, which included a civil DH Puss, a Fox, Bristol Bulldog fighters, a Hawker Fury fighter, a Tiger Moth (the best known of the Moths), a Hawker Hart day bomber, a Handley Page HP 42 civil transport, and the HP.38 (the prototype of the Hampden bomber).

The restrictions on photography were forgotten the following year. At the inaugural event, the morning had been devoted to an inspection of the aircraft and the afternoon saw a flying display, but in June 1933 the procedure was reversed. The afternoon was devoted to passenger flights and an inspection of machines on the ground. This was seen as a better arrangement; the visitor (and potential customer) would watch the machines fly during the morning and then pick those that had particularly impressed them, inspecting them in the afternoon.

The firms represented at the second SBAC had drawn lots for sequence of starting, and Handley Page Ltd, by a curious coincidence, drew No. 1 for the Heyford. The first machine to be demonstrated at the 1932 SBAC display was the

A single-seat fighter, the Bristol Bulldog first flew in 1927. In addition to Finland, the type was sold to seven foreign air forces, including the US Navy. (*Wings*)

The HP Heyford four-seat heavy night-bomber claims its place in the annals of British aviation history as the RAF's last biplane bomber. (*Wings*)

twin R-R Kestrel-engined prototype of the Heyford express night bomber. Sqn Ldr England took the Heyford into the air after a very short run and proceeded to put it through its paces. The speed range appeared large, and, in spite of the high position of the centre of gravity, the aircraft appeared both stable and controllable in flight. The landing was very good.

Mr Percival came next with one of his Napier Javelin-powered Gulls. Included in his repertoire were loops, long, steep climbs, and slow-flying demonstrations.

A de Havilland Dragon with two Gipsy Major engines (piloted by Capt. Broad) and a Tiger Moth with one (piloted by Mr Christopher Clarkson) took off 'in formation' and undertook a long, straight climb before going into 'synchronised loops'. Re-forming, the two machines flew along in formation, the Tiger Moth inverted and the twin-engined Dragon the right way up. For the rest of the allotted five minutes, Mr Clarkson undertook slow rolls and inverted flying—manoeuvres in which he was a past master.

The Saro Cloud, with two Armstrong Siddeley Serval engines, was piloted by Mr S. D. Scott and gave a favourable impression by its very steady flying. A peculiar slow-flying sideslip was a manoeuvre that had not been seen performed by an amphibian before.

Following the three Clouds, Flt Lt C. F. Unwins took the new Mercury-engined Bristol Bulldog IV up and gave one of his usual quiet and very 'finished' demonstrations of rolling, looping, fast and slow flying. The new version certainly appeared much faster than the service Bulldogs and quite as manoeuvrable.

The Hawker 'Super Fury' (or Fury II, serial K3586), piloted by Mr Bulman, with its Rolls-Royce Kestrel engine, shot across the aerodrome at 250 mph and disappeared in a long steep climb. It reappeared, and for the next three or four minutes Mr Bulman delighted the spectators by demonstrating the Super Fury's terrific speed, upward rolls, power dives, and S-turns. The 'High Speed Fury' became the Merlin engine test-bed.

Hawker also displayed its smart Hart demonstrator (registration G-ABMR), also with an R-R Kestrel engine. Piloted by Mr Sayer, it did some excellently executed loops, slow rolls, and climbing rolls, but his superb demonstration, good as it was, was inevitably overshadowed by the terrific power of the 'Super Fury'.

Armstrong Whitworth Aircraft Ltd were represented by a considerable number of machines and elected to entrust the flying of the AW 16 day and night fighter (registration G-ACCD) to Mr Turner Hughes, who gave a fine display. The machine was obviously very fast, benefitting from the high-speed Panther engine. The upward rolls and long, steep climbs terminating in loops, were particularly impressive. A slow flying demonstration gave some indication of the very wide speed range of this machine.

Two Blackburn Aircraft machines were demonstrated next—the Ripon torpedo-plane, flown by Mr Blake, and the B2 trainer, piloted by Mr Bailey. The Pegasus-powered Ripon appeared to handle remarkably well for a machine of this size and weight, and the Gipsy III-powered B2 excelled in slow rolls and inverted flying.

Above: The Tiger Moth is probably the best-known de Havilland Moth. It was used by the RAF during the Second World War for *ab initio* training of student service pilots. (*Wings*)

Below: The de Havilland Dragon biplane was the first light transport used by the RAF. This aeroplane belonged to No. 502 Squadron; it is now part of the Shuttleworth Collection at Old Warden, near Biggleswade, Bedfordshire. (*Wings*)

Below: The Hawker Hart biplane bomber was the RAF's premier type in service in the early 1930s, giving rise to the similar Hind, which followed it into RAF Bomber Command on its formation in 1936. Around 1,000 Harts were built in Britain, with a further forty-two built under licence in Sweden. (*Wings*)

The highlight of the show was when Supermarine's chief test pilot, Mr 'Mutt' Summers, looped a new single-engined Seagull V Walrus amphibian. The machine had flown for the first time off Southampton Water on 21 June, landed on its retractable wheeled undercarriage at Hamble, and later proceeded to Hendon. The flight at SBAC on Monday was only the machine's third flight. In spite of this, Mr Summers flew it as if he had practised for weeks. The machine seemed fast and was evidently manoeuvrable. The Bristol Pegasus pusher-mounted engine afforded the pilot an excellent field of vision.

The Pegasus-powered Vickers Wildebeest and Victoria were flown by Mr Banting and Mr Louis respectively. The former machine flew without its torpedo, and therefore its slow-flying routine was not quite as convincing as if a full load had been carried. The Victoria (serial K2808) appeared much faster, with two new 622-hp Bristol Pegasus engines, than the older versions, which had two 570-hp Napier Lions.

General Aircraft had entered the latest Monospar machine for the display, but at one time there was considerable doubt as to whether it would fly. However, the aircraft was cleared by the Air Ministry's airworthiness department following an early Monday morning test flight. The twin-engined Pobjoy-powered ST.6 (which resembled the earlier ST.4 but had an extra seat and retractable undercarriage) could show what it could do; with its raised wheels, it was very fast.

Avro's Mailplane, which had an Armstrong-Siddeley Tiger engine fitted, was flown by Mr Campbell Orde, the chief test pilot of Armstrong Whitworth Aircraft. It reached a speed of over 200 mph to challenge the Boulton & Paul biplane for the honour of being Great Britain's fastest mailplane. Mr Brown, Avro's test pilot, demonstrated the Avro Tutor. Mr Brown's demonstration of inverted flying on the Tutor was masterly. The aircraft's controllability was amazing, and Mr Brown demonstrated this by rocking it from side to side while inverted. The Tutor had a very impressive initial rate of climb (while inverted) of 500 feet per minute.

Two Spartan aeroplanes took to the skies next. They were the Spartan Cruiser medium-sized airliner (with three DH Gipsy Major engines) and the new little Clipper two-seater cabin (with a single Pobjoy engine). The Cruiser was piloted by Flt Lt Lynchblosse and the Clipper was piloted by Lt Col. L. Strange. The Clipper had just been to A&AEE Martlesham Heath, near Ipswich, Suffolk, for its official tests. Both aeroplanes were well-handled, but the two pilots confined themselves to normal flying that showed the aircraft very well.

The Bristol Mercury-engined Gloster day and night fighter, Type SS.19B, was excellently flown by Mr Saint, whose long, steep climbs, vertically-banked S-turns, and slow-flying demonstrations were very good. The lateral control at low speed was shown by rocking the machine from side to side.

Mr Christopher Staniland, Fairey's chief test pilot, displayed a Belgian-built Rolls-Royce Kestrel-engined Firefly. A quick succession of power dives, zooms, slow rolls at very low height, long vertical climbs, upward rolls, and inverted flying made for a very polished display. Fairey Aviation UK also showed its high-winged Fairey long-range monoplane, which in February 1933 had flown 5,309 miles in

The superb Avro Type 621 Tutor was the aeroplane that established the Avro Company as one of Britain's leading aircraft manufacturers. (*Author*)

fifty-seven hours and twenty-five minutes. Sqn Ldr O. R. Gayford DFC AFC was the pilot and Flt Lt G. E. Nicholetts AFC was navigator, and their efforts captured the FIA world's long-distance record for Britain.

Britain also secured the FIA world altitude record. Flt Lt C. F. Unwins, flying a Vickers Vespa, ascended to a height of 13,404 meters (43,975 feet).

Westland's Wallace two-seat general-purpose aircraft was flown by Mr Penrose, whose series of manoeuvres included some particularly noticeable zooming turns. The same company's new Wessex tri-motor, with three seven-cylinder Genet Majors, was also demonstrated by Mr Penrose. The Wessex was a new transport that had recently been delivered to Imperial Airways, and Mr Penrose's demonstrations included flying and landing with the central engine stopped.

The wide adaptability of Westland general-purpose aircraft had already been shown by the flights of two Westland aircraft over Mount Everest, made by the Houston-Mt Everest Flight. The aircraft was chosen by the Expedition Technical Committee out of the many other types considered due to its 'special purpose' performance.

Following the flying demonstrations in the morning, the guests were welcomed to luncheon in the huge tent by Mr Herbert J. Thomas of The Bristol Aeroplane Co. Ltd, Chairman of the SBAC, with some 1,200 guests being accommodated, all being carried out without a hitch. Following the toast to HM the King and the kings and presidents of the nations present, the only other speech was by Mr Thomas, who presided. After reflecting on the success of the previous year's SBAC flying display, he remarked that he hoped they would even improve upon their programme year by year. It gave attendees the opportunity to see the aircraft under the best conditions, in the air, rather than under just static conditions. For those who were directly interested and sought more detailed information, visits could be arranged to the various constructors' works, not only viewing the high-class material used but also the meticulous care with which the whole construction (under Air Ministry supervision) was carried out.

Mr Thomas emphasised the fact that both the personnel and materiel of the Royal Air Force would bear comparison with any other air force in the world. He was glad to note that not one country of importance was unrepresented that day. Moreover, they all appeared to use British aircraft or engines, instruments or materials. The British industry had steadily built up a reputation for quality in design, construction, and workmanship resulting in high performance. The display was the result of research programmes at the constructors' establishments, assisted by the experienced technical departments at the Air Ministry. An important factor in the progress of the industry was meeting the requirements of the British Air Ministry and of British commercial services, which demanded machines that were able to fly under widely divergent conditions. This resulted in the securing of many world records; the knowledge and experience thus acquired was reflected in the aircraft and engines that had been produced for other purposes.

In conclusion, Mr Thomas sincerely thanked the RAF staff, who were controlling the display that day, and also the Air Ministry for their generous cooperation in permitting the Hendon aerodrome to be used for the air display of their machines, which had contributed so hugely to the day's success.

A very graceful finish to the luncheon was then afforded by Mr Thomas, in consideration of the presence of many foreign visitors, who repeated his main remarks in fluent French. In all probability, the gathering included the most massed assembly of representatives of the countries of the world practically interested in aviation that had been brought together thus far. After lunch, the guests spent the afternoon partly inspecting the aeroplanes on the ground (or as passengers in them in short flights) and partly in the static exhibition that had been arranged in the old Airco sheds adjoining the Collindale Avenue entrance.

In addition to the aircraft already detailed that took part in the flying demonstration in the morning, the following firms and aircraft were represented in the static exhibition:

Armstrong-Whitworth Aurora 4 Imperial Airways
Blackburn M1/30A torpedo-bomber
Blackburn Commercial monoplane
Bristol Bulldog TM two-seater
Bristol Bulldog Type 120
Bristol and Vickers altitude machines
DH Fox Moth
DH Leopard Moth (flew over during the day)
Fairey Seal
Fairey Fox II
Fairey Long Range Monoplane

Handley Page Gugnunc
Hawker Nimrod

The following were in the aircraft park:

Hawker Horsley
General Aircraft Monospar ST.4
Avro Type 626 (shown with its fuselage panels removed to show the space for equipment)
Avro Club Cadet
Avro three-seater Cadet
Saunders-Roe Saro
Vickers Aviation Ltd M1/30 torpedo-bomber
Westland Wapiti

In the hangars, the following aero-engine firms were represented:

Armstrong Siddeley Motors Ltd showed one of the new Tiger engines and a seven-cylinder Genet Major.

The Bristol Aeroplane Co. Ltd was represented by a sectioned Pegasus running and Mercury.

Cirrus-Hermes showed a Hermes IV (with the new electric starter) and a Cirrus Hermes II.

The de Havilland Co. was represented by a Gipsy Major and a number of associated components.

D Napier & Sons Ltd showed a Rapier II and a Javelin.

Rolls-Royce Ltd attracted a great deal of interest with its sectioned Kestrel.

Numerous other aircraft firms and sub-contractors were represented, exhibiting parts, materials, and accessories.

The last SBAC to be held at RAF Hendon was in 1935 and featured the prototype DH 86 airliner and Avro Anson. The DH 86 was developed as a passenger aircraft for Imperial Airways and was to serve the military well in many roles over a long career—from trainer, communications aircraft, and light transport to bomber and reconnaissance aeroplane. Also shown was one of five DH Comet racers (civil registration G-ACSS) made famous by Scott and Campbell Black, winning the speed section of the 1934 England–Melbourne Air Race. It wore military serial K5084 as it was under evaluation by the RAF.

The last SBAC at Hendon also saw the last of Britain's biplane fighters, the Gloster Gladiator prototype. It had an 840-hp Bristol Mercury engine. It is interesting to note that of the thirty-eight aeroplanes displayed at the 1935 show, seventeen were biplanes, nineteen were monoplanes, one was an autogyro rotorcraft, and one was a pterodactyl.

Left: This DH Comet racer (registration G-ACSS) appeared at the last SBAC Hendon show in 1935. It was under evaluation for use in the RAF at the time. (*Wings*)

Below: The Avro Anson light transport prototype flew at the 1935 SBAC. This T.21 variant was produced many years later for RAF Flying Training Command. (*Wings*)

2

SBAC Hatfield 1936–1937

In 1935 it was decided that the SBAC show had outgrown the airfield and facilities at Hendon, and in 1936 the show moved to de Havilland's airfield at Hatfield, Hertfordshire. The show was separated from the RAF display but also exhibited a number of new types, including prototypes of the Supermarine Spitfire and Hawker Hurricane monoplane fighters, the Westland Lysander high-winged monoplane Army cooperation aircraft (famous for its special operations flights in Europe during the Second World War), the Fairey Battle monoplane light bomber, the HP Hampden bomber (which made some of the first raids on Berlin in August 1940), and the Bristol 142 (which was soon to be developed as the Blenheim). The fifth show staged at Hatfield was the last time it was restricted to a single day.

The 1937 show lasted two whole days, the first being devoted to the static exhibition and the second to the flying. Though entry was still by invitation only and foreigners were now banned—probably due to the unrest in Germany—it was proposed that the general public might actually be admitted in 1938. The public were already paying for most of the exhibits present through taxation.

The final show prior to the Second World War was held at Hatfield in 1937. Unsurprisingly, it featured a large number of military types, including the Miles Kestrel (which was developed into one of the most popular RAF advanced trainers), the Master, the Henley light bomber (adapted as a target tug), the two-seat Fairey P.4/34 (developed into the Fleet Air Arm's first monoplane fighter), the Fairey Fulmar, the Bristol Blenheim 1 twin-engined light bomber, and the twin-engined Airspeed Oxford (displayed in canary yellow Training Command colours; the Oxford would become an excellent multi-engine trainer). Civil types included the superb all-wood, twenty-two-seat DH 92 Albatross passenger airliner and an Imperial Airways Shorts C-class flying boat (which made a flypast).

The situation in Europe continued to deteriorate during 1938. The RAF was already well into its pre-war expansion plans and most SBAC members were concentrating on ramping up their production and developing new types to meet RAF requirements. The show was not held that year.

Britain's first monoplane fighter, the Hawker Hurricane 1, appeared at the Hatfield SBAC. (*Postcard*)

This Shuttleworth Trust Miles Magister (registration G-AJDR) is seen carrying military serial P6382. It is the only flying example of the Second World War training plane. (*Flight Library*)

3

SBAC Radlett 1946–1947

The first post-war SBAC was held in 1946 and relocated to the Handley Page aerodrome at Radlett, near St Albans, Hertfordshire, and attracted visitors from fifty countries. It was held at Radlett again in 1947 and extended to three days, with a fourth, non-flying day for technicians. Whereas all the pre-war events had been held in late June or early July, the shows after the war were held in the first half of September. This continued until 1968, when the organisers decided that the third week in September would be best; not unexpectedly, the show was a washout.

It was at Radlett that the British production jet engine was seen for the first time by the public. A visiting Soviet delegation were allowed to take an example back to Russia with them to be copied and reverse-engineered to power the Mikoyan design bureau's first mass-produced jet fighter, the Mikoyan MiG-15.

A great deal had happened during the SBAC's eight-year absence, and six of the sixty-one aircraft at the display were jet-powered (or partially jet-powered). Amongst them was the Avro Lancastrian test-bed, which had two R-R Nenes in its outer nacelles. A number of turboprops (essentially a jet engine with a propeller) were also under test on a number of different airframes.

Pure jets included a DH Vampire, a DH Sea Vampire, a DH 108 swept-wing experimental aeroplane, a Gloster Meteor IV, and a Supermarine E.10/44 prototype naval Supermarine Attacker, the Fleet Air Arm's first jet. Two Avro Lancasters converted as tankers gave a demonstration of in-flight refuelling, and a civil Airspeed Consul was displayed (a six-passenger post-war conversion of surplus Airspeed Oxford airframes).

At the still-private 1947 Radlett show, a number of interesting prototypes were displayed, most of which never entered series production. This included the single Percival P.48 Merganser, a small, twin-engined light transport with a high wing. The aircraft could carry up to five passengers; it first flew in 1947 and preceded a larger and more powerful version, the Percival P.50 Prince, which made its debut on 13 May 1948. When compared to its predecessor, the Prince feeder-liner retained

The DH Vampire F.3 was licence-manufactured in France by SNCASE as the SE535 Mistral. (*Warplanes collection*)

the same general twin-engined, high-wing structure, but it was considerably larger in size. The type later gave rise to the Hunting Percival Sea Prince military version, which was ordered by the Royal Navy as a communications aeroplane and used by the RAF as the Pembroke light transport throughout the 1960s and '70s.

4

SBAC Farnborough: 1948 Onwards

In 1948, the show moved to Farnborough, near Aldershot, Hampshire, south of London. For the first time, the general public were admitted on the final two days; the event now ran for six days, including three trade days and one technician's day. The first show at Farnborough also established the modern format that continues to this day, except that the show went biennial from 1962. Exciting new products from the post-war British aviation industry included: the Fairey Gyrodyne helicopter; the Rolls-Royce Nene, the first British jet engine approved for civil use, fitted in a Vickers Viking test-bed; the Vickers Viscount with Rolls-Royce Darts—the first four-engined turboprop airliner; the Armstrong Siddeley Mamba-powered Avro Athena (the first turboprop training aircraft); and the BP Balliol. The Avro Athena and BP Balliol both eventually turned to the Rolls-Royce Merlin piston engine for their power, and the Balliol became the RAF's standard basic trainer.

An interesting and unusual exhibit was that of the two Armstrong Whitworth AW.52 tailless research planes, one with Rolls-Royce Derwents and the other with Nenes. Even more unusual was the Planet Satellite, a four-seat touring aircraft with a V-tail, tricycle undercarriage, and a mid-fuselage engine driving a pusher propeller behind the tail assembly. Somewhat ahead of its time, this aircraft never flew, though it was rumoured it did make several 'hops' at Blackbushe Airport.

The 1949 show saw the first public appearance of the world's first jetliner, the DH 106 Comet prototype, which had made its maiden flight from Hatfield on 27 July 1949. There were a number of medium-size turboprop civil transports, including the Handley Page V, the improved Vickers Viscount, the unsuccessful thirty-seat AW 55 Apollo (powered by four AW Mambas), and HP Marathon 2, which made its maiden flight as a prototype on 23 July 1949. The Marathon 2 was a Marathon 1 that had been modified by the installation of two AS Mamba engines, each producing 1,010 hp. Similarly, the turboprop HP Hermes V was a direct development of the radial-engined Hermes 4. Unfortunately, a forced landing on the hills near Chilbolton resulted in one of the two prototypes being written off. With a top speed of 350 mph, the Hermes V was the world's fastest airliner at the time; however, it did not enter production.

The massive, eight-engined Bristol Type 167 Brabazon 1 (registration G-AGPW) flew over from its home at Filton, Bristol, for a flypast. It did not land, although it appeared at the show's static park the following year, by which time the second prototype (the Brabazon Mk II) was under construction. In February 1952, the government issued a statement that work on the Brabazon II would be temporarily suspended 'until more favourable conditions occur[ed]'. It later became obvious that the right conditions would never occur, and the cancellation of the entire programme on 9 July 1953 was no surprise. By that point, the Brabazon 1 (which had flown for the first time on 20 September 1952) had amassed a total of 382 hours and fifteen minutes in the air.

Two new military types appeared—the Avro Shackleton maritime-patrol aircraft (developed from the wartime Avro Lancaster bomber) and the English Electric Canberra (Britain's first jet bomber).

An Armstrong Whitworth AW.52 tailless research plane at the head of a varied field of British aircraft at Farnborough, 1948.

5

The New Jet Designs

By 1950, both Hawker and Vickers Supermarine were showing their new jet fighter designs—Hawker with its P.1040 and Naval N.7/46 in 1948, and Vickers with its Type 510 (which was basically a swept-wing development of its earlier straight-winged Attacker) in 1949. The Type 510 was the Royal Navy's first jet fighter, but it never fired its guns in anger. Hawker's N.7/46—also a straight-winged design—was developed into the Sea Hawk and provided the main fighter strength for the Royal Navy for six years; the type saw action during the Suez Crisis in 1956. It was developed through swept-wing versions as the P.1052 and P.1081, but neither entered series production. The latter, serial VX279, crashed near the Farnborough Airfield when Sqn Ldr T. S. 'Wimpey' Wade DFC AFC appeared to lose control in the landing circuit and ejected. Unfortunately, the ejector seat (not a Martin-Baker type) failed, and Wade lost his life.

The P.1081 ultimately led to the P.1067 Hawker Hunter prototype; the Supermarine 510 passed through a similar evolution as the Type 535, with a nose wheel as opposed to the Attacker's tail wheel. The Type 541 would become the Swift. Both the P.1081 and Type 535 appeared at the 1950 Farnborough show.

The show also featured competition for a new RAF trainer to replace the Percival Prentice, which had entered service in 1947. The Handley Page HPR.2 and Percival P.40 Provost both sought to secure the position, with the Provost eventually winning the day. The Provost went on to serve the RAF from October 1953 to October 1969. This basic piston-engined aircraft was developed into the Jet Provost and the ground-attack Strikemaster, with around 150 of the latter built for nine overseas air forces.

At the 1951 show, over half of the thirty-six flying aircraft and thirteen in the static display were jet-powered. The death of 'Wimpey' Wade determined that Hawker Siddeley Group chief test pilot Neville Duke flew the swept-wing P.1052 and also the P.1067 (serial WB188)—the Hunter prototype. The Hunter prototype had only flown for the first time two months earlier, on 20 July, but from the start, no major problems were experienced with this sleek, classic fighter. By the time of the Farnborough show, it was already approved for flying up to its maximum speed of Mach 0.94; at SBAC Farnborough, the main emphasis was naturally its speed. At this stage, pilots at the show were allowed to fly as low and as near to the crowd as they liked, although there were no advantages to getting too close. Many display routines are better viewed from a distance.

Although he had only flown the prototype Hunter for just a matter of weeks, Duke was well-practised in aerobatics in other Hawker types, including the P.1052 Seahawk and Fury. In line with other company test pilots, he had taken the opportunity to hone his skills with an impromptu aerobatic session flown over Langley at the completion of a test flight. In those days, no one complained about the noise associated with such flying, and as Langley was a company airfield, the airspace above it was unrestricted. As a result of this, he was quite polished in aerobatics with the new jets, and it did not take him long to perfect his Farnborough routine on the Hunter prototype.

The decision as to what to include in the display routine was left to the company and its pilot, and this was naturally dictated by what particular aspects of the aircraft's performance they wanted to emphasise and how much time you they been allocated for the routine by the show's organising committee. A five-minute slot was usual in those days, and this began at 'wheels rolling' and ended when the aircraft touched down.

Display flying for pilots at Farnborough can mean arriving ready for a week of high excitement in the cockpit, and often periods of sheer boredom on the ground. Once on site, the pilot tries to stay away from the throng of salesmen, PR people, press, and the public on the other side of the airfield. Segregation was almost complete in the pilot's enclosure, which comprised a white marquee surrounded by a high iron fence. The aviators briefed, ate, swapped lies regarding the others latest mount, and generally waited like gladiators to enter the main arena and perform.

Pilots normally assembled each morning at 11 a.m. for the day's weather briefing and flying display programme. In addition the demonstration sequence, radio call-signs and frequencies, time allowances, take-off and landing directions, taxi instructions, and parking and dispersal arrangements were detailed by the briefing officer. The essential time check and synchronisation of watches followed, as the success of the show relied on strict timekeeping. As many new flying restrictions were introduced at this time, these were reviewed—in particular as to how they related to the pilot, and to his displaying his aircraft in the best manner possible.

The most obviously anxious pilots at these meetings were usually the fighter 'jocks', their worry occasioned by the fact that their type of aeroplane normally carried minimal fuel but also a very large and thirsty engine. The pilot's fuel state was of even greater concern at Farnborough, as in order to save weight, each participating fighter had only a minimum amount of fuel on board—just sufficient for the demonstration, with a small amount to spare.

Following the briefing, the pilots dispersed to return to the pilots' tent for a splendid buffet lunch. Around ten minutes before their scheduled take-off slot, most would walk to their aircraft, which would be parked near the famous black sheds in those days. Prior to this, they would have donned their flying overalls and pulled on the 'bone domes' that were by now standard headgear for fast jet pilots. After a quick walk around the aircraft, the pilot would climb aboard and conduct his pre-flight checks before firing up his engine.

Once in the air, time literally flies by at Farnborough. Once the wheels are up, the aircraft usually curve away towards RAF Odiham (one of the diversion bases) in preparation for a 180-degree turn at full throttle for a high-speed run back along the runway. By this time, the controller would come over the R/T with, for example, 'Hunter, three minutes to go.' At this point, the most fast jet pilots would not be able to reply due to the effects of a 6g turn and speed building up towards around 700 mph, cramming his head down into his abdomen. Having lined up on the lead-in lights and fought off the blinding effects of 'black-out', the pilot would transmit, for example, 'Hunter coming in now,' over his R/T. At this moment, the show's commentator would wind the crowd up into a breathless state of anticipation as the as-yet-unseen fast jet approached them: '...at very near the speed of sound, and now available for export.'

Having completed the manoeuvre at the end of the run, the aircraft is pulled vertically into a 7-g climb before rolling onto its back as the speed falls, pulling through into line with the runway again. Although the display has only just begun, the controller's voice proclaims, 'Hunter, two minutes to go,' as the aircraft enters into a series of very high rate rolls, each of which take less than three seconds. Having re-engaged his eyeballs after completing a roll in the opposite direction, hauling round into position for a loop, the pilot takes an inverted position over the airfield. The voice over the R/T comes again—'Hunter, finish. Comet, you are clear for take-off.' That means the pilot has to get down quickly, replying to the controller, 'Hunter, on finals.'

It was then a matter of 'wheels and flaps down', getting the aircraft onto the final approach for landing. Afterwards, the pilots mostly felt as if they had undertaken something extremely strenuous for a long time. With a week of displays to fly and a need to remain fit and alert, it was important for the pilots to exercise restraint with respect to evening activities and social diversions.

Although attending Farnborough with the sole flying prototype of a new aircraft could affect the type's test programme, the impression left by the aircraft and its display pilot on both the public and prospective customers could play a great part in the future sales and success of the aircraft.

The first of the RAF's V-bomber trio, the Vickers Valiant, made its appearance in 1951 in an elegant silver finish. Somewhat less elegant was one of the two Short Sperrin bombers that was intended to be built as an insurance against V-bomber failures. The Sperrin ultimately did not enter series production, but it aided development and proved to be a useful test platform, including for the DH Gyron engine. Also shown for the first time in 1951 was the first de Havilland production Comet 1, which would prove disastrous for the company.

A number of new military prototypes surfaced at the 1952 show, including the Avro Vulcan delta-winged nuclear bomber, the RAF Transport Command Bristol Britannia 'Whispering Giant' airliner (powered by four 4,400-ehp Bristol Siddeley Proteus turboprops), the Gloster Javelin (the RAF's first custom-built all-weather night fighter), and the Fairey Gannet search and strike aircraft for the Royal Navy. The Gannet used an Armstrong Siddeley Double Mamba engine coupled to two contra-rotating propellers; it was subsequently developed into the Gannet AEW.3. All the above aircraft entered series production, but the massive Saunders-Roe flying boat passenger transport (registration G-ALUN), which was powered by ten Bristol Proteus gas turbines in coupled pairs and two singles, did not.

Over the next few years, the show featured its fair share of production and prototype aircraft, plus a number of flying test-beds. These included the Gloster Meteor fighters and English Electric Canberra medium bomber. The latter was the first foreign type to be licence-produced in America, being renamed as the Martin B.57 and seeing service in south-east Asia—although in truth, in addition to its slightly enlarged size, the B.57 was almost a completely different aeroplane.

By this time, it was possible for the regular visitor to the SBAC to judge how the post-Second World War British aerospace industry was facing up to its new challenges.

In 1953, the third V-bomber made its appearance—the crescent-winged Handley Page Victor—in addition to an impressive gloss-black Vickers Valiant 2, but the latter was not developed.

The 1954 show featured a brilliant display by the diminutive Folland Midge lightweight jet fighter, which was developed into the Gnat dual-seat trainer. The Midge gave a foretaste of what would come ten years later, when the Yellow Jacks aerobatic team (forerunner to the Red Arrows) used the same mount when becoming the RAF's official aerobatic display team. The Midge was replaced in this capacity by the BAe Hawk T.1 in 1980. Other display teams of the era included the Hawker Hunter-equipped 'Blue Diamonds' of No. 92 Squadron and the 'Black Arrows' of No. 111 Squadron, the Jet Provost-equipped 'Flying Pelicans' from CFS Cranwell, and, later, the English Electric Lightning-equipped No. 74 'Tiger' Squadron, which signalled the end of the use of front-line types for display flying.

Anyone who attended the 1954 show will surely remember the amazing sight of the large Avro Vulcan Type 698 bomber being rolled by test pilot Wg Cdr 'Roly' Falk, in addition to the superb demonstration by Jan Zurakowski in the big Avro Canada CF-100 fighter, which culminated in a 'falling leaf'. In contrast, the year also marked the first appearance of the lumbering Blackburn Beverley transport for the RAF, which had been developed from the earlier General Aircraft Universal Freighter; the type was notoriously bad to fly in turbulent air, when it would drop like a stone. The author experienced this during a horrendous flight from Bahrain to Sharjah in 1964.

XR760

Above: A camouflaged long-range Avro Vulcan B.2A with a tail-cone ECM fit lands at Farnborough. (*Take Off*)

Opposite above: The EE Canberra B.2 was the RAF's first jet bomber. The aircraft served the RAF in many roles over a thirty-year period—not least in high-altitude reconnaissance. In the new millennium, No. 39 (PR) Squadron has flown the armed Predator UAV in Afghanistan. (*Author*)

Opposite middle: An HP Victor at RAF Marham, Norfolk, following its conversion to a K.2 in-flight refuelling tanker. (*Author*)

Opposite below: BAC (EE) Lightning F.3—the RAF's first supersonic air-defence fighter. The type was never 'used in anger' by the service. (*Wings*)

There was a plethora of new types at the 1956 show, including; the ultimate Supermarine Swift development; the FR.7, which superseded the FR.5 and usurped the disappointing F.3 fighter; the much-improved Hawker Hunter F.6 day fighter; the Navy Fairey Firefly, used as a fighter, reconnaissance anti-submarine, and trainer aircraft, with total production amounting to 1,570 airframes (including fifty-four to the Royal Netherlands Naval Air Service and several overseas customers); the Fairey Gannet AS.4; the DH 106 Comet C.2 for RAF Transport Command; the EE Canberra B(I).8 interdictor bomber for RAF Germany; the Scottish Aviation Twin Pioneer light transport—the first RAF aircraft to carry air-to-air missiles; the DH Vampire T.11 trainer; the DH Venom NF.3 night fighter; and the DH Sea Venom FAW.21. Despite this, 1958 would be the year when the number of service types participating (totalling 149) would exceed the number of other types in the show.

The first British truly supersonic type was seen by 1957, by way of the English Electric P.1A and B-model precursors to the famous EE Lightning air defence fighters. As the number of export contracts increased, the aircraft were displayed in their respective air force markings or airline liveries. For example, a Bristol Britannia appeared in El Al colours, a DH Vampire T.11 trainer in those of the Lebanon, a Folland Gnat for Finland, a Percival Pembroke for the West German Luftwaffe, and a Vickers Viscount for KLM.

The Blackburn NA.39 prototype was the world's first specially designed, low-level, high-speed maritime strike bomber, and it appeared at the show in 1958. It was later named the 'Buccaneer'. That same year, the RAF Black Arrows from No. 111 Squadron and their colleagues displayed and looped twenty-two Hawker Hunters at Farnborough. Easily the most advanced design displayed at the show was the forty-eight-seat Fairey Rotodyne, which had first been flown in 1956. The Rotodyne was a vertical-take-off airliner with two conventional Napier Eland turboprops for forward flight at a cruise speed of 160 knots, enabling city-centre to city-centre operations. The Rotodyne was cancelled in 1962, but it was replicated in the twenty-first century by the nineteen-seat VTOL Bell/Boeing BA.609 tilt-rotor based on the design of the military Bell/Boeing V.22 Osprey. Also on display at the 1958 show was the Shorts SC.1 experimental aircraft, which made the first public demonstration of Vertical Take-off and Landing (VTOL); unfortunately, its display was brought to a premature end by grass ingestion, although the aircraft was able to show what it could do in the 1960 show.

The Bell/Boeing BA.609 tilt-rotor uses similar technology to the military V-22 Osprey. It is aimed at the nineteen-seat commercial market. (*Air Power International*)

6

Ready at Last to Accept Overseas Friends

The new president of the SBAC, Sir George Dowty, spoke at the Society's annual dinner in London on 7 September 1960. During the speech, he mentioned the following salient points:

> As a representative of the equipment industry I am proud of the success of this side of British aviation, which perhaps tends to lose its identity in an industry primarily thought of in terms of complete aircraft or aero-engines.

However, Sir George warned:

> ...we may have to face a decline in our aviation exports however, for everywhere we saw a downturn in orders for military aeroplanes ... it [is] fortunate that at this time we [are] looking to expand our civil business, to meet an increasing demand for people and goods to be transported more quickly with greater economy. There are many matters to be resolved before we could start a supersonic airliner project. So far in the field of air-freighting we have only scratched the surface, a field where great opportunities lie, with only 5 per cent of the country's total exports being air-freighted to their destinations.

Sir George also referred to the new structure of the industry and thanked Mr Peter Thorneycroft, Minister of Aviation, for the hospitality afforded to the SBAC the previous evening and for the keen interest the government had shown in SBAC affairs. Mr Thorneycroft assured the members that the British aircraft industry was assured of the government's continued support and that 'some monies would continue to trickle through'.

Of spaceflight, the minister said:

> For the United Kingdom alone to seek to compete in the field of manned space vehicles with Russia and the United States would certainly be imprudent and

> probably impossible. There are, however, spheres of space research, commercial and scientific, which fall far short of this. If such research and development, whether in the field of satellites or launchers, is to be carried on outside the United States and Russia, let us make an effort in the Commonwealth and in Europe to find a co-operative basis for it. If we are to look towards the stars let us look a little beyond our own national frontiers first of all.

Of supersonic transports, the minister said:

> There is apparently a limited market for supersonic commercial aircraft. If one of us goes into it, we may well all attempt it, and I am bound to say that I believe, in this country, we would do it very well. There is, however, in this supersonic field, a powerful case for all of us agreeing what to make, when to make it and then making it together. Our studies of a supersonic civil aircraft are indeed well advanced and will soon be advanced still further. We are very ready to consider how we might collaborate with the Americans, the French, or other overseas friends who are interested.

When Sir George Dowty, the president of the SBAC, spoke on Sunday morning, on the last day of the twenty-first SBAC, he spoke of an industry being 'so full it [couldn't] find enough workers', adding that he could see the situation continuing for the next two or three years. He also said that the Short Brothers and Harland SC.1 was the undoubted star of the 1960 show. As to the future of these shows, the SBAC Council was to discuss whether to include overseas products incorporating British components; however, an international exhibition on the lines of the Paris Air Show was not contemplated. Over the course of the show, orders totalling £20 million were received for transport aircraft. These included:

Five DH Comet 4Cs
Sixteen AW.660 Argosys for RAF Transport Command
Six Handley Page Heralds for Jersey Airlines
Three Avro 74s for BKS Air Transport
A 'substantial number' of Westland P.531/2 Mk 1 helicopters for the British Army

There was also a potential order for twelve Avro 748s from Greece, together with a licence to manufacture.

Attendances at the Farnborough '60 show continued the upward trend observed before 1959, when the long, hot summer brought an artificial peak. There were 22,000 people at Farnborough on Friday 9 September, the first of the three public days; on Saturday, the attendance climbed to 100,000, and on Sunday it climbed even more (as did the temperature, which reached 22 degrees Celsius.

Dan Air operated five DH Comet 4C airliners until the airline was sold to British Airways for a nominal £1. (*Author*)

The first of the trade days, Tuesday 6 September, was cloudy with intermittent rain, and all but the well-protected refrained from even a brisk walk through the static park. Instead, they stayed within the confines of the exhibition tent or the terraced enclosures, occasionally making a quick dash from one to the other.

Inside the 'big tent' there were some 300 stands, from the very smallest to the 'big two'. On the Hawker Siddeley stand, the products of the member companies seemed thoroughly integrated— that is to say, all mixed up, with Avro 748, Armstrong Whitworth Argosy, communications satellite, and attendant salesmen all in the one area.

In contrast, the British Aircraft Corporation had divided its area neatly into quarters, placing a separate company in each, and co-ordinated the four with a 10-foot piece of advertising backdrop.

Despite the gloom and steady rain, the flying display opened at 2.30 p.m. with the Bomber Command 'scramble', this time by Avro Vulcan B.1As of No. 617 'Dambusters' Squadron. The first aircraft was moving after forty seconds and in the air after one minute and ten seconds, and all four were airborne in one minute and forty-seven seconds—nine seconds faster than the Vickers Valiant had managed the previous day.

Avro Shackleton MR.2 'L' for Lima of No. 201 Squadron RAF Coastal Command touched down smoothly on the main runway after a twenty-two-hour flight that had begun at Farnborough during Monday's display. AVM L. W. C.

Bower (AOC No. 19 Group) and Wg Cdr Alan C. Davies (Squadron Commander of No. 201 Squadron) took regular turns at the controls. Their mission had taken them to Madeira, a point off Lisbon, out over the Atlantic, where they had spoken by radio to a French weather ship, and back to the English Channel for a routine training exercise. The area covered was approximately 80,000 square miles and the distance flown was some 3,700 nautical miles. Their average height was between 800 and 1,000 feet.

The show continued well despite the appalling weather, with the Shorts SC.1 experimental VTOL aircraft blowing up clouds of spray from the runway as it hovered before and after its quick sprint circuit. Two newcomers made their appearance on the first trade day. The wingspan of the Westland Rotodyne had been considerably increased from 18.5 feet to 75 feet, with the fuselage extended by 4 feet and a number of other changes, including a redesigned rotor head, tip-jets, cockpit canopy, and fins and rudders, with the third central fin now standard. The other newcomer was the military AW Argosy 'Whistling Wheelbarrow' transport, which flew over in two runs with its rear-loading clamshell doors alternately open and closed. Unfortunately, the amount of surface water on the runway prevented the two RAF Lightning interceptors from taking part.

For the twenty Hawker Hunters which took part in the show, it was a triumph. Hawker test pilot Bill Bedford in the Hunter 66A was limited in what he could do by the weather, but David Lockspeiser stole the show with the lowest and fastest run of the day in the Royal Navy FR.10 variant. For 'Treble One' and the 'Black Arrows', these were the worst conditions they had ever been asked to perform in, and yet they

Following the cancellation of the HS Nimrod AEW 3, the No. 8 Squadron AEW.2 Avro Shackleton served to 'fill the gap' until the Boeing E.3 Sentry became available. (*Wings*)

excelled, tailoring their display to the wet and hazy visibility above Farnborough. In between the opening circuit—performed in place of the loop— and the final double-loop, for which all eighteen aircraft were together, the main display was flown by the nine-man section led by Sqn Ldr Peter Latham, with the second section of nine aircraft out of sight, off-stage, until the finale. The show's commentator remarked that the 'Draken formation .... [bore] some resemblance' to the Saab all-weather fighter.

Although still damp, on Wednesday the 7th the rain held off throughout the day and the outdoor exhibits received their fair share of attention. The various missile demonstrations were most popular. On one site, the RAF showed its airfield defence Bloodhound SAM and launch controller in action. 'The launch controller presses the firing button...' the recorded commentary related, at which time the eyes of many of the onlookers were fixed on the aircraft itself. Needless to say, nothing happened this time. A little further along, the British Army 'D' Troop of 60 Guided Weapons Battery, 36 Guided Weapons Regiment, Royal Artillery, was busy showing off its Thunderbird. It was the first unit to be equipped with the weapon, but at that time they had not yet actually fired it. Not to be left out, the Navy demonstrated a typical firing sequence for its Seacat aboard a 'typical' destroyer.

Again the flying display began with a Bomber Command 'scramble', this time by the HP Victors of No. XV Squadron, which were the focus of attention at precisely 2.30 p.m. The scramble time for the first aircraft airborne was one minute and twenty-seven seconds, and all four were away in two minutes and eight seconds.

The programme followed the planned schedule and it was the agile Westland helicopter 'circus' that jigged along the runway, with the Rotodyne retiring into the background after a curtain-raising run. The Avro Vulcan bomber once again appeared to leap into the air, momentarily appearing to stand still before climbing vertically at almost an impossible angle. Treble One's Hawker Hunters were again superb and Bill Bedford flew the Hunter 66A in a low high-speed pass. However, it was James Dell in the dual-seat BAC Lightning T.4 trainer who scattered the runway-fringing crowd with the fastest, nosiest run of the day at an estimated modest Mach 0.95. Next was the Westland 'circus'. [

Witnessing this superb spectacle were the West German Defence Minister, Herr Strauss; the US Secretary of the Air Force, Mr Dudley C. Sharp; and Soviet designers Oleg Antonov and Alexander Yakovlev.

Thursday 8 September arrived with the airfield covered in a thick, white morning mist, but the forecast indicated that the mist would lift by 10.30 a.m. and thereafter it would be a nice day. This proved to be basically accurate, although the timing was somewhat inaccurate. In the static park, Blackburn Aircraft were expounding the virtues of their new Buccaneer strike bomber by way of a bearded engineer in loud blue overalls handing out literature to visitors, while Vickers, in a similar manner, entrusted the same task to two very attractive TCA stewardesses. Not unexpectedly, the queue for the Vanguard airliner was much longer than that for the Buccaneer.

Above left: The Westland Gazelle HC.1 was used by the Army Air Corps for helicopter-pilot training at Middle Wallop, Hampshire, until the formation of the JHF Flying Training School at RAF Shawbury. (*Author*)

Above right: The Westland Wessex HC.2s of No. 60 (R) Squadron were the last of the type to serve with the RAF. They served at RAF Benson, Oxfordshire, from June 1992 until its disbandment on 1 April 1997. (*Author*)

The Westland Wasp light anti-submarine strike helicopter was operated from platforms on small ships. (*Author*)

Since 2000, Royal Navy Westland Sea King AEW.2s are all upgraded AEW 7 variants. (*Author*)

The Westland Sea King HC 6 (Sikorsky S.61A) is the Royal Navy's premier anti-submarine warfare (ASW) type. (*Author*)

Westland's AH.9 Super Lynx is the upgraded version of the company's venerable AH.7 strike-attack helicopter. (*Author*)

By midday it was quite warm and there was little doubt the flying display would start on schedule. The Vickers Valiant of No. 148 Squadron Bomber Command, which had opened the show on Monday, returned to try to beat their original 'scramble' time of one minute and fifty-six seconds, which they did in no mean fashion. The first aircraft took about the same time as on Monday to become airborne-—one minute and six seconds—but the intervals were closed up and all four were airborne in only one minute and thirty-seven seconds.

The rest of the display continued in time-honoured Farnborough fashion, the RAF's Hawker Hunters and four Jet Provost T.3s of the CFS providing the best publicity for the service. Bill Bedford expanded his thrilling export Hunter 66A display to fifteen turns in his smoke-trailed spin, while Keith Isherwood made his Farnborough debut on the EE Lightning F.1 supersonic air defence fighter.

A major announcement on this day was that Rolls-Royce and Hispano-Suiza had concluded a licence agreement for the joint production of Tyne turboprops.

On the first of the public days, Friday 9th, things continued very much as before. A couple of newcomers were added: the RAF Gloster Javelin FAW.4, doing 'Derry turns'; two HS (Blackburn) Buccaneers in close formation, one of them later showing the DayGlo interior of its bomb bay, then both landing from opposite directions; four EE Lightnings of No. 74 'Tiger' Squadron, led by Sqn Ldr John Howe; No. 111 Squadron, with its eighteen all-black Hawker Hunters performing as one and in two impeccable patterns of nine; and, finally, an RAF Avro Shackleton maritime-patrol plane departing on its twenty-two-hour NATO exercise.

Saturday's display followed in a similar vein, except that the Lightning T4 landed accompanied by the acrid smell of burning rubber due to the failure of a Maxaret unit. Wg Cdr Dell skilfully managed to keep the aircraft straight and taxi off the runway to a stop. The port wheel was changed before the T4 resumed its journey to dispersal.

As Saturday's crowd of 100,000 began to go home at 4 p.m., No. 148 Squadron Vickers Valiants from RAF Marham, Norfolk, began arriving for Sunday's V-bomber scramble. The pilots were totally oblivious to the fact that the 'ban-the-bomb' campaigners had lined the road from Farnborough railway station with their billboards, which carried various anti-nuclear slogans.

Earlier, Tom Brooke-Smith had impressed with his display of the Short SC.1, the HP Herald demonstrated its under-600-yard departure with thirty men of the 1st Parachute Regiment Support Company on board, the Westland 'circus' had given their impressive rotary-winged-craft demonstration, and the RAF's Victor and Vulcan bombers displayed their massive power output during their climb outs. The four Victors of No. XV Squadron completed the Bomber Command scramble in one minute and forty-five seconds.

Sunday saw blue skies and excellent weather. The display was poignant in that Short SC.1 display pilot Tom Brooke-Smith landed to conclude his test-flying career after greatly contributing to the development of VTOL and the Hawker P.1127 Kestrel, which was still to be test-flown at that time. Brooke-Smith's display was impeccable, transitioning from forward to vertical flight in front of the president's tent.

Another final appearance involved the 'Black Arrows' of No. 111 Squadron, who gave their last performance as the RAF's official aerobatic team. When he came back on air after the RAF commentator had finished describing the display, the show's commentator, Oliver Stewart, rightly pointed out that the skill and judgement needed for the spectacular stream landing by eighteen Hawker Hunters was as great as that required for their airborne manoeuvres.

Although it was probably overlooked by most visitors, the 1960 show featured the first public exhibition of a genuine example of British space hardware forming part of an active programme. Following discussions between the British and American governments, several research groups in the UK were developing instruments to be placed in orbit by American Scout rockets beginning in 1961. Scout 1 was planned to carry five experiments recommended by the Royal Society's National Committee on Space Research, all for the investigation of the ionosphere or associated solar parameters.

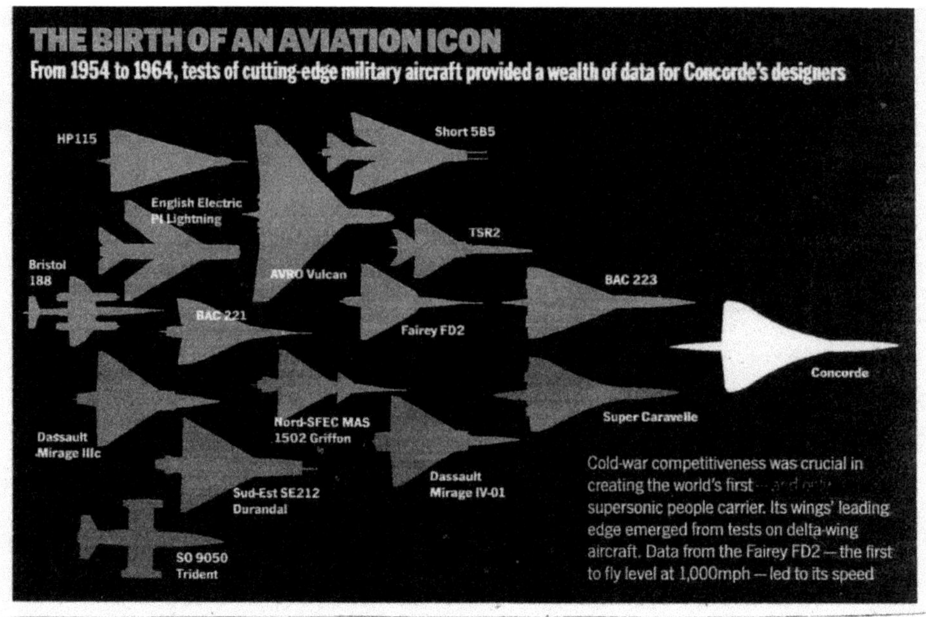

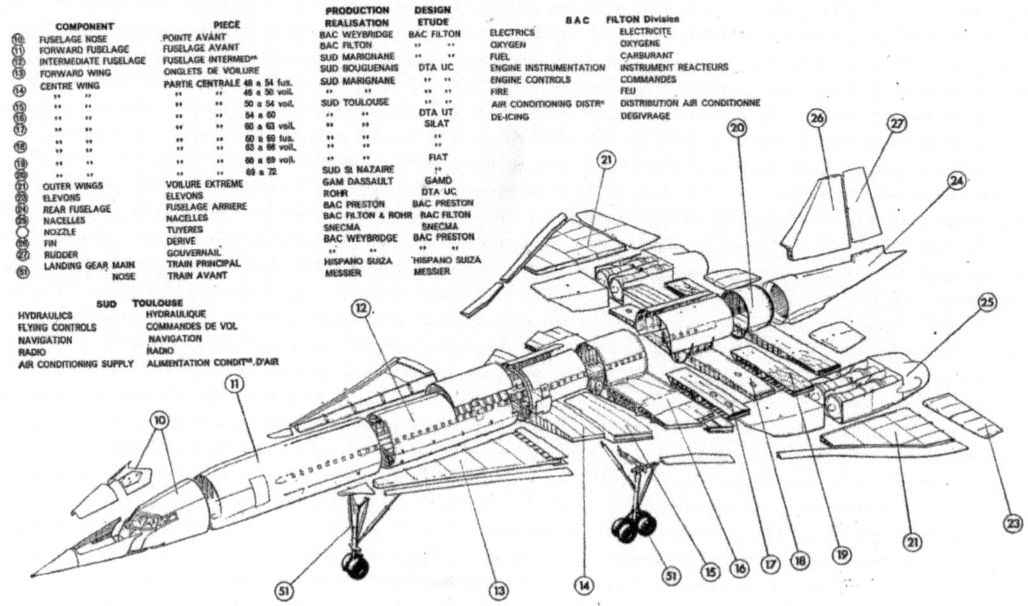

The birth of an aviation icon—Concorde. (*Wings*)

7

Research Planes: Concorde, The Birth of an Aviation Icon

Farnborough 1961 was remembered more for the new types that did not appear rather than those that did. A number of new and interesting aeroplanes had been completed by the British industry during the previous year, but many were unable to make their debut at the 1961 show for various reasons.

Perhaps for this reason, the twenty-second show accented on Army aviation. Not only was it the first occasion at which the Army made an individual contribution to the flying display programme (with their mirror-image manoeuvres of two formations each of four Saro Skeeters), but the service was also given precedence over the other two in order of appearance. Among the brand new aircraft was Beagle Auster's Mark 11 AOP, which had a zero-wind take-off run of less than 100 yards and could land in less than 70. Four Army Westland Skeeter helicopters were prominent, as were the parachutists of the SAS, who undertook a spectacular drop from 7,000 feet, free-falling to 5,000 feet before opening their parachutes. The RAF's No. 1 Parachute Training School at Abingdon, Oxfordshire, also demonstrated free-fall during the display.

The September show was significant in that it was the 100th flying display staged by the SBAC. At the first show, at Hendon in 1932, there were thirty-four aircraft; at the 1961 show, there were approximately 150, though most were in-service military types, with manufacturers demonstrating some twenty-three different types. Only two of these were entirely new—the Beagle B.206 and the Handley Page HP 115 experimental slim-delta, which had made its maiden flight at RAE Bedford on 17 August.

Four types of missile were on show for the first time in Britain—the Ferranti Blue Streak, the de Havilland Red Top air-to-air missile, the EE Blue Water tactical bombardment missile, and the RAE Jaguar, a joint development with the Australian WRE that RAF Sepecat Jaguars had test-fired at the Woomera Range in the north of southern Australia. In addition, the stand-off nuclear Blue Steel weapon developed for the RAF V-bombers was shown on its loading trolley in the missile park. Among the 8,500 invited visitors expected at the show were the Russian designers Ilyushin, Yakovlev, and Antonov.

Other interesting exhibits included the yellow-and-red Gloster Meteor F.8 (serial WH505) developed by Flight Refuelling as a pilotless U.16 target drone, the de Havilland Engines' all-blue Gyron Junior DGJ.10-powered Gloster Javelin FAW.1 test-bed (serial XA552), the black Napier icing-research Avro Lincoln (registration G-29-1) and the de Havilland Comet 4C of Middle East Airlines, with its pleasing, large cedar-of-Lebanon insignia on the tail. The research Avro Lincoln carried a HS (Blackburn) Buccaneer tail-plane 'bullet' section (added to series Buccaneer aircraft to prevent buffeting) and was earmarked later to test, in turn, a 'supersonic fighter intake', a cargo-aircraft wing section, a submarine-search aerial system, and a laminar-flow wing section. Under its No. 4 nacelle was a section of Napier all-bonded helicopter blade undergoing endurance testing.

The Hawker P.1127's flight development had been going steadily and successfully since October 1960, and it had explored both ends of its flight envelope, from hovering up to bursts of sustained level flight, from high subsonic speeds to the stall. However, at this time, the all-important transition from the hover to forward flight had not been made, and the company questioned whether its presentation to visitors and potential customers without this achievement would be worthwhile. The development of the aircraft was inextricably linked to the new Bristol Siddeley Pegasus BS.53 vectored thrust engine that made it all possible, with all the attendant work and complications that this involved.

This same scenario prevailed with regards to the hypersonic Bristol 188 research aircraft and the Handley Page H.P.115 (serial XP841) slim delta for exploration at the lower end of the speed scale. The all-steel Bristol design represented a major advance in structures as well as aerodynamics, being intended for sustained flight at Mach numbers well above Mach 2.0, to investigate the effects of kinetic heating on the airframe, engines, and accessories. The power plants in the first prototype were the 14,000-lb static-thrust DH Gyron Junior DGJ 10R reheat turbojets, although the nacelles were designed for the rapid installation of other engines in the research programme. In its original form, the Bristol 188 was intended to have been used for investigations into armaments and other propulsion systems, though it was subsequently only used for structural and aerodynamic research. This narrow-delta research aircraft had completed only seven and a half hours of flying before appearing at the show.

Many of the results achieved by the Bristol 188 research plane were applicable to the development of the Anglo-French Concorde SST, as were those associated with the Fairey Delta 2, the Fairey Company's last fixed-wing type, which raised the FIA world speed record to 1,132 mph—310 mph above the previous record set by a North American F-100 Super Sabre. The FD.2 was the first British aircraft to exceed 1,000 mph.

In the civil field, de Havilland had rolled out its Trident airliner prototype (registration G-ARPA), in BEA colours, at Hatfield in August. The long-nosed, clean-winged jetliner—the first British example to have its trio of Rolls-Royce Spey power plants grouped at the tail—was not due to fly until the end of the year. Even so, twenty-four had already been ordered by BEA, with an option on a further twelve.

Another type to 'miss' the 1961 Farnborough was the Hunting ER.189D research plane, which was of a somewhat-odd configuration with its high, long wing, high-set tail-plane, fixed tricycle undercarriage, and single-seat cockpit. It was powered by a Bristol Orpheus turbojet, of which almost all the thrust was ejected through slits along the wing surfaces and over full-span, double-slotted flaps. The system claimed to give exceptionally high flight coefficients and low-speed performance and control, and also to reduce drag.

Those types that did feature at Farnborough '61 included:

Armstrong Whitworth AW. 660 Argosy C.1

The military version of the civil 650. The RAF had ordered fifty-six for Transport Command. The aircraft primarily differed from its civil counterpart in having clam-shell or 'crocodile-jaw' rear-loading doors. In RAF service, the Argosy was nicknamed 'The Whistling Wheelbarrow' due to its twin-boom tail section and the high-pitched whistle emanating from its four 2,680-hp Rolls-Royce Dart 101 (R.Da.8) engines.

Avro Vulcan B.2 Bomber

This aircraft was developed to carry the Avro/HSA Blue Steel 100A stand-off nuclear missile recessed under its fuselage, with the bomber's bomb bay doors specially modified to permit this.

Avro 748 Series 100

This aircraft was a forty-to-forty-four-seat feeder liner to be produced in two versions—the Series 100 and Series 200. The former was powered by two 1,740-hp R-R Dart 514 turboprops and the latter by two 2,105-hp R-R Dart 531 engines.

Beagle-Auster A.109 Airedale

The first in a range of new British light planes, the original prototype (registration G-ARKE) was displayed at Farnborough with the R-R/Continental 260-hp engine.

English Electric Lightning T.4 (Serial XM974)

The side-by-side dual-seat T.4 trainer differed from the F.1 fighter by having a larger cockpit, though the two aircraft were otherwise identical. Instructor and pupil were seated side-by-side as in the Jet Provost, and as the T.4 was fully equipped with Ferranti Airpass radar and the twin Firestreak armament pack, operational training sorties could be carried out.

Folland Gnat T.1 (Serial XM693)

The Gnat entered service with RAF Flying Training Command to replace the DH T.11 Vampire as its standard advanced trainer. A pre-production batch of fourteen aeroplanes and an initial production batch of thirty were ordered from 1958 to 1960; this was followed by a further order for another twenty in July 1961 and a final contract in March 1962 to bring the total RAF deliveries to 105, with series production aircraft produced by Hawker Siddeley.

Hawker Hunter T.66

The T.66 was followed by the 66A and B versions. The former was a two-seat trainer supplied to the Indian Air Force (IAF). The Hunter 66A was the Hawker Company's demonstrator and similar to the IAF aircraft, with only minor modifications. The 66B was also of dual-seat side-by-side configuration and was for the Jordanian Air Force. All were developed from the original RAF dual-seat Hawker Hunter T.7 trainer.

Hunting Jet Provost T.4

The T.4 was a development of the earlier T.3 and started life as a private venture. It eventually supplemented the T.3 in RAF Flying Training Command, to be complemented by the T.3A and 4A and eventually the pressurised BAC T.5/5A.

Short SC.1

Originally known as the PD.11, the SC.1 was an experimental aircraft commissioned for research into the field of vertical take-off and landing, associated with the development of the Hawker P.1127 Kestrel. Two SC.1 were built to be used for the ongoing development test programme at this time.

Hawker's Hunter T.66 demonstrator T.66A was supplied to the Indian Air Force. (*Author*)

The dual-seat Hawker Hunter T.7 Raspberry Ripple was used by the EPTS at Boscombe Down to teach pilots 'spin recovery'. In the new millennium, the Tranche 3 Saab JAS-39 Gripen (leased by the hour) is used. (*Author*)

Handley Page Victor B.2 (Serial XL164)

Pictures showing the RAF Handley Page Victor nuclear bombers carrying a Blue Steel missile could be seen in the missile park.

Hawker Hunter 66A (Registration G-APUX)

The Hawker Company exhibited its demonstrator for the dual-seat aeroplane for the export market, with 230-gallon underwing tanks, to show its ability to act as a navigation trainer, possibly for the Lockheed F-104G Starfighter specially developed for the West German Luftwaffe. It was claimed that the full F-104G cockpit and attack system could be accommodated. G-APUX was later used by the Royal Aircraft Establishment to trial Rank Cintel head-up display instrumentation.

Civil Types at Farnborough '61

Avro 748 Series 1 (Registration G-ARMV)

The third 748 produced (the seventh including structural test aircraft and assemblies), later destroyed by fire. It was displayed in UK independent Skyways colours as it was to be used for six weeks of route-proving after the show.

DH Comet 4C

The fifth series production aircraft for Lebanon Middle East Airlines but the sixtieth Comet 4 produced. Five examples would be built in Chester, in the north of England, for RAF Transport Command.

Handley Page Herald Series 200 (Registration G-ARTC)

The first Series 200 Herald to be publicly displayed, the aircraft was a lengthened version of Handley Page's feeder liner. It appeared in the colours of Maritime Central Airways, who, together with their associate Nordair, had ordered three.

Vickers VC 10 and Super VC 10

These were displayed on British Aircraft Corporation's stand to enable a comparison between the standard and lengthened variants.

Beagle B.206 (Registration G-ARRM)

Among a range of Beagle types and models, this was probably one of the most exciting new aeroplanes at the show. It was the first 'big' Beagle, with two Rolls-Royce-built American 260-hp Continental engines.

The DH Comet 4C was ordered by Argentine Airlines. (*Wings*)

De Havilland Dove 8 (Registration G-ARDH)
A standard production aeroplane with two Gipsy Queen 70 Mk 3 engines; it was used as a company demonstrator.

De Havilland 125-100
A Rolls-Royce Viper-powered executive jet. Two models—one part cut away—appeared on the Hawker Siddeley Aviation stand.

Helicopters at Farnborough '61

Westland (Bristol) Belvedere (Serial XG459)
A standard aircraft from the RAF production line, fitted with nineteen troop seats.

Westland Rotodyne (Serial XE521)
A Westland development of the Fairey Rotodyne experimental VTOL transport was due to fly over from White Waltham daily to take part in the flying display. A large model on the Westland stand demonstrated the main differences from the Fairey aircraft abandoned in February 1961.

Westland Wasp (Registration G-WASR) and Scout (Serial XP165)
G-WASR was a five-seat passenger version of a Nimbus-powered Wasp; it was displayed in company colours on the Westland stand. Serial XP165 was a military production Scout that appeared in the flying display each day.

The RAF used DH Dove light transports as communications aircraft. (*Wings*)

DH (HS) 125-100 biz-jets were adopted by the RAF Dominie T.1 navigation trainers to replace the elderly Vickers Varsity at No. 7 FTS at RAF Stradishall, near Newmarket, Suffolk. (*Author*)

Westland Whirlwind (Registration G-APDY)

A civil Series 3 Gnome Whirlwind demonstrator shown at the twenty-second SBAC show, fitted with eight seats and finished in company colours.

Westland Wessex Mk 1 (Serial XM300)

A Royal Navy machine equipped with dipping-sonar equipment, Doppler aerials, and minesweeping gear. On each side of the fuselage were stub wings to which homing torpedoes or extra fuel tanks could be fitted.

Gloster Javelin XA552 after it had been modified to flight test the de Havilland Gyron Junior engines. It carried a metal nose with a pilot boom and was finished in overall blue with white markings.

8

The Show Goes Biennial

'The best show we have had'—that was how Sir Roy Dobson, SBAC President, summed up the Society's twenty-third flying display and exhibition, which took place in 1962 and drew a record attendance of 330,000 people. Sir Roy confirmed that the next show, which would take place in 1964 unless something sensational happened in the meantime, would continue to show British products or those licence-manufactured in the UK.

Among the VIPs that attended on the first Trade Day, Tuesday 4 September, was newly appointed Minister of Aviation Mr Julian Amery, who revealed that the government had decided in principle to order a number of DH 125 executive jets for use as navigational trainers for the RAF. The Minister was asked about the SST project. His only comment was that 'he looked forward to meeting his French opposite number to discuss this'.

As usual, when the Jet Provosts of the CFS made their formation take-off to begin the flying display, it was raining—ironically, there had just been a prolonged bright spell while everyone had lunch. The early items on the programme—the Bristol T.188, the HP 115, the Vulcan Olympus engine test-bed, and the CFS and BAC Jet Provosts—did well in spite of the appalling conditions. Nevertheless, the sky was blue again for the Royal Navy's contribution. Of note were the 'twinkle roll' by all four of the Supermarine Scimitar quartet; the superb 'rising sun' burst by Fred's five in their DH Sea Vixens; and the final 'follow-my-leader' low pass by the five Supermarine Scimitars, which produced four audible near-sonic bangs. That was the flying completed for the day.

Continuing a naval theme, during the day, Westland announced that 'a substantial quantity' of Westland Wessex 5 helicopters had been ordered by the Royal Navy for Royal Marine commando duties. Westland also announced that it had reached an agreement with Autair Helicopters Ltd of Montreal for the 'joint operational exploitation' of Westland Hovercraft in Canada, and that a version of the SR.N2 as used for the Channel crossings would be taken to Canada for testing under Canadian conditions.

The following day, Wednesday 5 September, low cloud-base precluded loops by the RAF Hunters and Lightnings, but their incredible take-off—eighteen Hunters plus eight Lightnings—was as spectacular and traumatic as ever for those near the runway. As the Lightnings took off, using reheat and twin jet pipes housing balls of fire, the noise rose to an ear-shattering crescendo.

The take-off of the Hunters and Lightnings were undoubtedly the best for noise, but for breath-taking simplicity, impact, and wonder, the twin Hawker P.1127s were the highlight. These were aircraft to remember; at one minute they were flying elegantly at speed, and at the next they were hanging in the air like helicopters, as if on an invisible line, clearly demonstrating the meaning and implications of the VTOL fighters.

An order from the Venezuelan Air Force for fifteen Hunting Jet Provosts was announced by BAC during the afternoon. Delivery of the aircraft, plus spares and ground equipment, was to begin in late 1962. The aircraft was to be an armed version of the Jet Provost Mk 4 trainer, which was already in service with the RAF. Four other air forces had already adopted the Jet Provost for ground-attack duties, including Saudi Arabia, Sri Lanka, Sudan, and Kuwait.

On Thursday 6 September, the forecast was for the rain and thick drizzle to clear, and in fact, in the morning the rain did clear to enable visitors to visit the static park and view those exhibits that had had their tarpaulin weather covers removed.

In the tents, as today, the latest innovations in advertising were being exercised to grasp the visitor's attention and concentrate one's thinking along company lines—

At the 1962 show, Hawkers chief test pilot Bill Dedford used his vintage display Hawker Hurricane as a 'taxi' to fly between Dunsfold Aerodrome, Surrey, and Farnborough, Hampshire. (*Warplanes collection*)

films, recorded panegyrics, illustrations, pictures, diagrams, brochures, hand-outs, public relations officers, and all the trade-show paraphernalia.

At the 12.30 p.m. press briefing, the Director of the RAE, Dr M. J. Lighthill, spoke about the 'aerospace plane' that had been mentioned briefly by Mr Amery at the SBAC dinner the previous evening—and mentioned rather less briefly in the morning's newspapers. At the same time, Shorts Brothers and Harland were waiting at enclosure C 9 to tell the press two things. First, Shorts had signed an agreement with Chance Vought, Hiller, and Ryan for co-operation on the XC-142 tilt-wing tri-service VTOL transport (should this aircraft be chosen in the NATO NBMR-4 competition). Secondly, the Tigercat land-based version of the Seacat ship-to-air missile was being offered to the Army.

The afternoon's flying display included only one P.1127, which was displayed in pouring rain. The second aircraft had returned to Dunsfold and its flight development programme. At the same time, BAC announced that the T.188 would take two days out to resume its flight development programme at Boscombe Down, but today it would hopefully fly its passes along the Farnborough runway. As on previous days, the Avro 748 turboprop short-haul airliner impressed with an ultra-short take-off. The Navy and RAF appeared with their usual dash and spectacle; sixteen Hawker Hunters from No. 92 'Blue Diamonds' Squadron replaced eighteen from No. 111 'Treble One' Squadron, flying a superb 'diamond'. It was the ultimate in design and motion, seen head-on, lazily rolling and, at the crest of a high loop, entering a smooth smoke-on loop, all against what by this time was a clear blue sky.

On Friday 7th, the first of the public days, the Soviet designer Andrei Tupolev did as *Flight International* had predicted the Russians might do due to lack of official invitations; he paid his £1 entrance fee at the gate like any other visitor and promptly asked to look over the BAC VC 10. His request was initially refused, but it was subsequently granted and he was finally entertained at lunch by BAC's Sir George Edwards. Eight RAF parachutists performed freefalls from an AW Argosy, landing expertly on their target. The FAA Officer's Association presented to leaders of the Royal Navy aerobatic teams.

Finally, in the late-summer-afternoon sun, there was a display of vintage aircraft that included the Fairey Swordfish, the DH Mosquito, and the RAF Se.5a—adding a distinct contrast to the modern types.

On Saturday, the 82,000 attendees who arrived to enjoy the show watched as a full vintage display was held as a living tribute to fifty years of British military aviation. This included displays by John Oliver in a Mosquito, Lt-Cdrs Spafford and Wreford in a Fulmar and Swordfish respectively, Peter Varley in a Gloster Gladiator, Duncan Simpson in a Hawker Hart, 'Willie' Williamson in a Bristol Fighter, Grp Capt. Pan Hanafin in an Se.5a, David Morgan in a Spitfire, and Bill Bedford in Hawker Hurricane. Bedford used his mount as a 'taxi' to return to Dunsfold to pick up the P.1127 Kestrel for his demonstration. Another feature of the day was the arrival at 2.30 p.m. of Westland WS-55 Whirlwind HAR.10 (serial XP400), flown by Sqn Ldr Leo De Vigne, after a successful 6,000-mile demonstration tour of Lebanon, Syria, and Egypt.

It was a cloudless day on Sunday, disproving the somewhat gloomy weather forecast. The day brought a crowd of over 110,000 to witness some of the week's best flying, including the P.1127 transferring from vertical to forward flight, then *vice versa* on landing, and then flying to dispersal instead of taxiing. Ranald Porteous aero-batted the Beagle 206X, and Lt Paul Millett RN pulled the HP115 around in dramatic, steep turns to show 'an exotic moth' in plan-form, the apparition accentuated by the large RAF roundels on its underside. The special Avro Vulcan Olympus engine test bed almost burnt up the runway, while four No. 700Z Flight Blackburn Buccaneers from the Intensive Flying Trials Unit at RNAS Lossiemouth, Scotland, took part in the Naval demonstration. A solo Hawker Hunter flown by Flt Lt Douglas Bridson undertook upward rolls to 21,000 feet and Nos 74 and 92 Squadrons rounded off a superb show with a 'diamond' made up of twenty-five Lightnings and Hunters.

In 1962, the first of the post-Second World War second-generation British jetliners appeared—the DH (HS) Trident with three Rolls-Royce Spey by-pass engines. Three HS Tridents appeared in the flying display, the type best-known for making the first automatic landing on a scheduled BEA service to Paris on 16 May 1967. Unfortunately, the Trident was a sales disappointment, with only 117 built. The BAC (Vickers) VC 10 was also a disappointment, with only thirty-nine built; the type had rear-mounted engines and was exceptionally quiet inside. It was specially commissioned for BOAC, who operated thirty of the aircraft—a mixture of standard and stretched 'Super' versions—at one time. Fourteen donned RAF livery, with a number of ex-airline Super versions later converted to air refuelling tankers.

Hawker displayed two V/STOL P.1127 Kestrel prototypes that had first flown on 21 October 1960. The Kestrel the forerunner to the HS Harrier GR 1, which first entered RAF service on 1 April 1969. De Havilland displayed its first executive jet, the DH 125, which subsequently achieved great success under the Hawker Siddeley and British Aerospace banner until sold to Raytheon in the United States in the 1990s.

By the time of the 1964 show, the SBAC had become the Society of British Aerospace Constructors, with the appearance of another new British airliner, the BAC 1-11. After a shaky start, the BAC 1-11 achieved a degree of export success—particularly in China—and, like the Vickers Viscount turboprop, managed to penetrate the difficult but important American market. With production totalling 242, sixteen BAC 1-11s fitted with hush-kits were still in use in 2006—four in Africa, five in Asia, and seven in Europe.

The Shorts Belfast freighter, which made use of the Bristol Britannia wings and tail design (Shorts had built a number of Britannia airliners under sub-contract),

The HS 125-400 continued to attract orders at the Farnborough '62 show. (*Author*)

Raytheon displayed its Hawker 800, although the new jet was essentially a development of the BAe 125. (*Author*)

also made its Farnborough debut in 1964. The ten examples built served for just over ten years (1966 onwards) with No.53 Squadron at RAF Brize Norton, Oxfordshire, until they were withdrawn from service in a round of defence cuts. A number entered the civil field, operating with the Heavy Lift cargo carrier for many years; ironically, they were subsequently used by the RAF for heavy freight transportation.

There was a radical change in policy for the show in 1966, when the exhibition of foreign aircraft was permitted as long as they used British engines or contained other major British components. The show was now less of a public spectacle and more of a serious trade show; the public attendance over the three days was down from 240,000 in 1964 to 180,900 in 1966. However, overseas guests on the trade days had risen from 9,000 to 11,000.

There was ample participation at Farnborough '66 by European types, including the Fokker Friendship, the Breguet Atlantic 1 maritime-patrol aircraft (powered by two Rolls-Royce Tynes) and the similarly powered Transall C-160 transport (the first prototype in *Armée de l'Air* colours), the Italian Agusta-Bell 204B helicopter, the Paggio PD 808, and the Macchi MB.326 jet trainer. However, the Fiat G.91T dual-seat fighter trainer had been the most successful. SBAC Director Sir Richard Smeeton said:

> The SBAC Council was to consider whether this principle could be applied in future to aircraft built in the USA and fitted with British engines and equipment, but there will have to be some rationing of the American effort.

The director predicted that the 1968 show would be a vintage one, particularly as the Hawker Siddeley HS 801 maritime Comet (Nimrod MR.1), BAC/Aerospatiale Concorde, and Sepecat Jaguar would all be flying.

Concorde's engine, the Bristol Siddeley/SNECMA Olympus 593, became airborne for the first time on Friday 9 September 1966 from Filton Airfield, Bristol, installed beneath the bomb bay of its Avro Vulcan B.1 test-bed aircraft. The following day, Saturday, the Olympus Vulcan (serial XA903) made its debut at the SBAC show, flying over with Bristol Siddeley's HS.125 in attendance after the RAF Red Arrows had completed the main public flying display. Bristol Siddeley's chief test pilot, Tom Frost—who earlier the same day had been flying the vintage Gloster Gladiator at Farnborough—reported that the '593' had lit perfectly and ran smoothly. The Olympus Vulcan made a return visit on the final day of the show, then departing to get on with the 250-hour programme of test flying planned for the '593'.

A fascinating feature of the public days at the '66 show was the lunchtime flying display by veteran and vintage aircraft. Neil Williams was particularly outstanding in his ultra-aerobatic Stampe. Sunday began wet, and families ate their picnics beneath the exhibit's 'big tent' instead of on the grass outside. The weather later dried up slightly, but cloud remained to hamper the main flying display.

Breguet Atlantic 1 maritime patrol planes were one of Europe's first collaborative ventures. It also saw extensive input from the USA aircraft industry; the aircraft were produced to meet a NATO requirement. (*Author*)

DH Comet 4 airliners were reconfigured by Hawker Siddeley to provide the RAF with the world's first and only maritime-patrol plane, the Nimrod MR.1. It was ultimately upgraded to MR.2 standard. (*Air Power International*)

The Anglo-French Concorde SST was a regular visitor to SBAC Farnborough.

On the first of the trade days, Tuesday 6 September, a distinguished selection of visitors included the Duke of Edinburgh (making his first visit to an SBAC show), a group of ministers from the Ministries of Aviation, Technology, Trade, and Defence, and a Soviet delegation led by three deputy ministers and including six designers.

In the morning, BAC and Elliott-Automation Ltd held a formal press conference in the exhibitors' restaurant to talk about automatic landing and the BAC VC 10 and 1-11 airliners. Hawker Siddeley Dynamics held an informal conference to give details of the firm's share of the global communications satellite contract.

The Ministry of Aviation announced that the NGTE rigid-lifting rotor would be demonstrated on a vehicle on the runway at precisely 2.15 p.m., but it just went along the runway instead of taking off.

A number of sailplanes and autogyros took off as curtain-raisers for the afternoon's main display. The stars of the main event were Capt. Orazio Benato and Capt. Riccardo Peracchi of the Italian Air Force, from the AMI Experimental Flight Centre at Pratica di Mare, near Rome. They had brought a dual-seat Fiat G.91T/1 (serial MM6432) and a Macchi MB.326 trainer (serial MM54183) powered by a Bristol Siddeley Viper engine. These were ordinary aeroplanes that were spectacularly flown with 'verve and panache', as the commentator declared. The display culminated in loops and rolls on approach.

The new Red Arrows (formerly the Yellow Jacks, but still equipped with Folland Gnats) were designated as the RAF's official premier display team in 1965. The flight at Farnborough on the Tuesday was the team's sixty-third display since the opening of the new season on 24 April 1966. The team was from Flying Training

Command's CFS and operated under the management of Sqn Ldr Dick Storer, with New Zealander Sqn Ldr Ray Hanna as team leader.

On Wednesday 7 September, BAC announced that United Air Lines had reserved six Concorde SSTs, bringing the total orders to date to sixty for thirteen airlines. The flying was curtailed by the non-participation of some of the stars, including the naval air group of Buccaneers and Sea Vixens from HMS *Hermes*, the Red Arrows, and the intrepid Italians. As a number of *ad hoc* flight demonstrations to potential customers continued, most of the guests were heading for the gates or back into the company enclosures or the 'big tent'.

Among the aviation enthusiasts who visited the show on Thursday were MP Edward Heath and the King of Thailand. Mr Heath found time to visit the press tent to say that Concorde was good for technology, but also that the 'air bus' had to be purely commercial.

Beagle announced the sale of six B.206s, netting around £300,000; the company also stated that over 100 people had committed to the B.121 Pup two-seat club trainer, a mock-up of which was displayed.

A BAC Lightning purposefully ran into a bed of gravel in a demonstration of the RAE's latest attempt to arrest an overshot aeroplane. However, the most important Lightning at the show was the F.53 with Arabic markings and 'Royal Saudi Air Force' on the fuselage, but bearing RAF military serial XR770. To all who viewed it, it was the identical twin of the Lightning F.6 standing next to it (serial XP697) in the static display. Both aircraft were fitted with belly tanks and exhibited various weapons that they could carry, including ground-attack armament such as two 1,000-lb bombs, Matra missiles, 68-mm SNEB rockets, and a forty-eight-cell glass-fibre rocket pack interchangeable with Firestreak and gun packs. Apart from the main contract, Saudi Arabia also received a number of Lightning F.52s (ex-RAF F.2s) and T.54s (ex-RAF T.4s).

Previewing the wing-shape of Concorde in miniature, the BAC 221 research aircraft flew in from RAE Bedford to make its first appearance at the show. Dart-like and vortex-tipped, the blue delta made fast and low runs, looking disconcertingly like Concorde as it banked round in the distance and reared up on its spindly landing gear, with needle-nose drooped, for an un-parachuted landing.

Less exotic types included the Fokker F-27 Friendship, which was flown superbly by Jas Moll and Peter Brey. There were also two PBN Islanders—the prototype, G-ATCT, and the first production aircraft, which appeared in the livery of Loganair. The HS 748 also appeared, as did the HP Herald (registration G-ASBP), which was later delivered to Air Manila, and a military BAC VC 10 from RAF Transport Command. HS Andover C.1 serial XS595 was also shown, finished in a two-tone brown with black undersides and a white cockpit roof patch.

A BAC 1-11 had flown over briefly on the first day from Wisley. By this time, the airliner had notched up 103 sales, with options for another twenty-three. An order for four from Channel Airways was announced on the opening day of the show.

The BAC (EE) Lightning, equipped with ventral 'belly tank', displayed by BAe in the static park. (*Author*)

Another show-stealer was the Handley Page HP 137 Jetstream, which had appeared in mock-up form and bore the registration G-ATXH. A small, high-speed executive transport with two Astazou turboprops, the type was later adapted as a small regional airliner to carry up to eighteen passengers. At the time of the Farnborough show, Handley Page had received orders for 120, with the type due to fly in the spring of 1967 and first deliveries to customers scheduled for 1968.

The following year saw the unfortunate demise of this famous British company, which fell victim to the government 'rationalisation' programme that sought to merge all the smaller aircraft firms into larger units. The independently minded Sir Frederick Page (1885–1962) held out for what he thought was a fair deal, only to see orders for the HP Victor bomber cut back and the potential RAF purchase of his Herald transport diverted to Avro. The Handley Page firm survived its founder by just seven years.

The earlier restrictions on participation in Farnborough were gradually relaxed until 1972, when European aerospace companies were allowed to exhibit without such restrictions. By 1974, the show had become truly international.

In 1966, attendees were already taking advantage of the gradual policy change. Two Italian Aermacchi MB.326 trainers were in attendance, as was NATO's Italian hope for a common European-produced fighter, the Fiat G.91. Also featured were the short-haul Fokker F-27 Friendship airliner, the Piaggio-Douglas PD-808, the Franco-German Transall C.160 military transport, and the Italian-built Agusta-Bell AB-204B helicopter. An English Electric Lightning was hastily repainted in the markings of the Royal Saudi Air Force (retaining its RAF serial) in confirmation of their order for forty of the aircraft. The first low-cost ten-seat high-wing STOL light utility and short-haul aircraft, the Britten-Norman BN-2 Islander, made its first appearance at this show. After many changes of ownership over the years, the type and the company have survived into the twenty-first century.

After the Second World War, the international and European collaboration programmes helped to save the UK's aviation industry from total collapse—despite their problems. Two major benefits arise from collaborative ventures; one is that it is more difficult to cancel these projects, mainly due to the intricacies of international agreements. The second is the fact that the size of initial orders are increased, and the work is shared proportionally dependent on the number of units to be produced.

Nevertheless, with very little in the way of a sound foundation at home, the ability of the British sales teams to secure overseas orders has been quite remarkable. It is of interest to examine both civil and military aircraft production in the immediate post-Second World War period, and in particular to note the percentage of export sales to home-based deliveries (usually around 30 per cent) of the total build.

The Franco-German Transall C-160 medium transport was another attempt from NATO to standardise equipment, but most members (including the United Kingdom) chose the ubiquitous Lockheed C-130 Hercules instead. (*Author*)

Military Aircraft

Period	Type	Total Production	No. Exported	% of Total Production
1945	Avro Lincoln	528	18 new, 12 refurbished	
1945–1950	Gloster Meteor	3,788	800 new, 133 refurbished	26.3
	DH Vampire	4,206	1,043 new, 108 refurbished	27.4
	Hawker Fury	1,079	354 new, 59 refurbished	38.2
	Percival Prentice 509	156 new, 1 refurbished	30.8	
	Percival Prince/Sea Prince	74	18 all new	
	Vickers Valetta	262		
	HP Hastings	152	all new	
1950–1955	Avro Shackleton	187	8 all new	
	Boulton Paul Balliol	222	5 new, 7 refurbished	
	Bristol Sycamore	177	65 new, 1 refurbished	
	DH Chipmunk	1,270+	274 new, 29 refurbished	23.8
	EE Canberra	1,376 in 22 variants	153 new, 42 refurbished	14.1
	DH Venom	840	102 new, 6 refurbished	12.9
	Hawker Sea Hawk	538	104 new, 32 refurbished	
	Westland Dragonfly	137	41 all new	
	Supermarine Attacker	181	36 all new	
	Supermarine Swift	174		
	Fairey Gannet	378	52 new, 18 refurbished	
	DH Sea Venom	294	39 all new	
	Westland Whirlwind	418	90 all new	
	Vickers Varsity	163	1 new	

1955-1960	Avro Vulcan	136		
	Blackburn Beverley	49		
	DH Sea Vixen	145		
	Gloster Javelin	428		
	HP Victor	104		
	Hunting/Percival Provost	461	88 new, 25 refurbished	24.5
	Hawker Hunter	2,322	480 new, 156 refurbished	27.4
	Hunting Jet Provost	459	65 all new	14.2
	Folland Gnat	145	40 all new	27.6
	Vickers Valiant	104		
	Supermarine Scimitar	76		
	Saunders Roe Skeeter	78	10 all new	
	Westland Widgeon	12	6 all new	
	Westland Wessex	352	27 all new	
1960	EE/BAC Lightning	388	54	13.9
	Blackburn Buccaneer	60	16 all new	
	BAC 145/167 Strikemaster	150	150	100
	Westland Scout	132	8 all new	
	Westland Wasp	177	79 all new	
	Westland Sioux	199		
	Hawker Siddeley P.1127	15		

Civil Aircraft

On the civil side, between the end of the Second World War and the closure of the BAe RJ line in 2003 at Woodford, Lancashire, the UK produced 6,033 civil aircraft, including the BAe 125 business jet. The total sales of HS/BAe/BAe Systems 146/Avro RJ series production aircraft failed to beat the 444 sales record set by the Vickers Viscount turboprop that was produced from 1948 to 1963. Of the remainder, most were built in only small numbers, with a few exports.

Period	Type	Capacity	Total Produced	No. Exported	% of Total Produced
1945–51	HP Hermes	40–80	29		
	Avro Tudor I	30–60	22		
	Avro Tudor II	79			
1945–1947	Vickers Viking	36	163	66	24.39
1945–1953	DH 104 Dove	8–11	542	409	75.5
1946–1951	Miles/HP Marathon	14–22	43		
	HP Marathon II	28			
1951	Bristol 170 Freighter	214	174		
1947–1953	Airspeed Ambassador	28–36	23		
1953	Scottish Aviation Pioneer	59	15		
1953	Hunting Pembroke	133	72		

1948–1963	Vickers Viscount	50–80	444	356	80.18
1949–1964	DH/HS Comet 106	50–100	113	35	30.9
1950–1963	DH 114 Heron DH 114 C.3/C4 Queens Flt VIP/ VVIP	14–17 1/2	143	103	72.0
1952–1960	Bristol Britannia	90–140	87	26	29.8
1955–1968	HP Herald-Dart Herald	50–60	56	48	85.7
1957	Scottish Aviation Twin Pin	88	33		37.5
1959–1962	Vickers/BAC	90–120	43	23	53.5
1960	Vickers Vanguard	44	23		52.3
1959–1966	AW/HS Argosy 650	65–71 freight (8 civil, 56 RAF)	7		9.8
1960–1988	Avro/HS/BAe 748/Andover	50	379	53	14.0
1962–1970	Vickers/ BAC VC10 BAC Super VC 10	100–150 15	47	6	12.8
1962–1978	DH/HS Trident	90–150	117	16	13.7

1962–1995	DH/HS/BAe 125	4–8	893 (UK built)	(new-build exports to 45 countries) (used aircraft to sixteen countries)
1963–1986	Shorts Skyvan SC 7	19	150	
1965	Britten-Norman (PBN)	9–11	1,220	
1967–1993	HP Scottish/BAe Jetstream J31/J32/J41/61/ATP 220/ 180+/106/2/63	571+		
1968–1971	Britten-Norman Trislande	16–18	85	
1969–1979	BAC/Aerospatiale Concorde SST	100–140	10 (UK built) 20 total	
1974–1992	Shorts SD330/360	33–39	139 or 164 including Shorts C-23A Sherpa (militarised SD-330/360 built originally specifically for USAF)	
1981–2003	DH/HS/BAe/BAe 146/Avro RJ BAe 146 209, Avro RJ 166			

Short's Skyvan light transport received some orders from third-world air forces. Its enlarged 'Sherpa' form (with stretched fuselage) was ordered by the USAF; the type is still in USAF use in the new millennium. (*Warplanes collection*)

A Soko Galeb G.2 ground-attack aircraft of the former FRY Air Force. (*Author*)

Farnborough 1968 saw the first appearance of Japan at the show, displaying its NAMC YS-11 twin turboprop transport used by Japanese regional airlines and the military. Also making their first appearance were the Yugoslavians, who exhibited their Soko Galeb and Jastreb trainer light ground-attack aircraft.

Shorts showed no less than five SC7 Skyvans, with more than 150 sold between 1963, when it made its maiden flight, and 1986. The aircraft featured an unusual bonded light-alloy construction, giving the airframe strength while maintaining an extremely low weight. The rear bottom panel hinged to form a rear-loading ramp. The ultimate development of this light freighter was the Shorts 360, *via* the 330 and military Sherpa (which was initially ordered by the USAFE to transport aero-engines and other spares and components in the European theatre, but was later transferred to the US Army.

Beagle displayed five pups and two B.206s; the latter was produced as a crew transport for Bomber command nuclear V-bomber crews to fly to dispersal airfields.

Sud showed its Puma tactical transport and Gazelle training and light-utility helicopters, demonstrating its recent association with Westland at Yeovil, in the West Country.

Perhaps the most important appearances of all at the 1968 show were that of the first series-production Hawker Siddeley Harriers for the RAF, which entered service on 1 April 1969.

The Hawker Siddeley Nimrod—the first jet maritime reconnaissance aircraft—at Farnborough Air Show in September 1968.

9

First Attendance by the USA

In 1970, a modest first attendance by the United States saw two operational US Navy Vought Corsairs from the USS *Kittyhawk* and a civil Gulfstream II. The Anglo-French Concorde SST prototype 002 was in the flying display only, but further proof of Anglo-French cooperation at this time was evident in the appearance of three Sepecat Jaguar GR.1 close air-support fighters. These fighters were finally withdrawn from front-line RAF service in 2006, after thirty years, in their upgraded GR.3/3A form. A civil BAC VC 10 test-bed was shown with its two port engines replaced by two Rolls-Royce RB-211 engines. At the 1972 show, the Lockheed TriStar—for which the RB-211 power plant was developed—was also shown.

The 1972 show was primarily remembered for the number of aerobatic display teams that participated, including: the RAF Red Arrows, with nine Folland Gnat T.1s; the eleven Fougas Magisters of the Patrouille de France; ten Fiat G.91 PANs of the Italian Frecce Tricolori; and six Westland Sioux (licence-produced Agusta-Bell 47G-3Bs) from the British Army's 'Blue Eagles' team. Saab displayed two 105 basic/advanced jet trainers that had first flown on 29 June 1963, a MFI-17 Supporter, and the superb double-delta Saab Viggen AJ37 (which superseded the J35 Draken with the Swedish Air Force). Germany and Australia also participated with the former, showing a Dornier Skyservant light transport, a VFW 614 airliner, a MBB BO105 helicopter, and an Australian Government Aircraft Factory (GAF) Nomad light transport. India sent a pair of licence-built HAL Ajeet/Gnat single-seat fighters and the indigenous Rolls-Royce Viper-engined HJT-16 Kiran jet trainer.

Opposite above: An Irish Air Corps Gulfstream IV. The Gulfstream II is used for coastal patrol and surveillance; the type was initially a surprise selection ahead of a rotary-winged type. (*Aviation News*)

Opposite below: The DRA Sepecat *Jaguar* T.2 Raspberry Ripple is used for experimental work at the UK Defence Research Establishment at Boscombe Down. (*Author*)

10

SBAC Farnborough Becomes Truly International

The year 1974 saw the SBAC display and exhibition at Farnborough become truly international for the first time. Of the ninety-three aeroplanes exhibited, twenty-seven were from the USA, nine French, seven Swedish, five German, four each from Italy and Poland, three from the Netherlands and Romania, and eight representing international collaborative ventures.

McDonnell-Douglas exhibited its F-15 Eagle—arguably the best jet fighter of the twentieth century, and certainly the one with the most 'kills', seeing combat both with the USAF and the Israeli IDF/AF and Saudi AF in the Middle East. However, it was Lockheed that was the most prominent overseas manufacturer, with its massive C-5 Galaxy strategic transport for the USAF, the P-3C Orion maritime-patrol plane and S-3A Viking for the US Navy, its Mach 3.0 SR-71A Blackbird high-altitude reconnaissance stealth plane (specifically designed for USAF CIA-sponsored covert operations over Russia and Eastern Europe), and the L-1011 TriStar airliner, which was powered by Rolls-Royce RB 211 turbofan engines that at one time, due to seemingly insurmountable technical problems, threatened to bankrupt the famous British engine maker.

The SR-71 Blackbird arrived in the rain on 'press day' after making a record-breaking flight from New York in one hour, fifty-five minutes, and forty-two seconds, flying at 1,807 mph. Aerobatic flying was confined to the Red Arrows' Gnats and, on the public days, the Rothmans aerobatic team (flying Pitts S-2As—a vintage group) and the Royal Navy's Historic Flight sole torpedo-equipped vintage Fairey Swordfish biplane.

Farnborough International '74 Participants

2–8 September 1974
Main Public Days: 7 and 8 September

The MDC F-15C Eagle is arguably the best jet fighter/interceptor of the twentieth century; its combat record both with the USAF and IDF/AF is second to none. (*Author*)

Lockheed's P-3 Orion has attended SBAC Farnborough on numerous occasions. Although it is an excellent aircraft, Lockheed has not secured any new orders for upgraded variants. In the new millennium, a number have become available on the second-hand market.

The Fiat G91 was to be NATO's first common fighter plane for all European member states, though this never materialised. The Italian AMI, the German Luftwaffe, and Portugal used the type. This example is a Luftwaffe G91R equipped with nose radar. (*Warplanes collection*)

Above left: The Mach 3.0+ Lockheed SR.71 Blackbird high-altitude reconnaissance plane is likely to remain the fastest aspirated aeroplane ever to fly. (*Aviation News*)

Above right: Having been withdrawn from use during the 1990s, the Lockheed SR 71 Blackbird was reinstated to front-line service by US President Bill Clinton for surveillance of the demilitarised zone between North and South Korea. It was also used over the Balkans, with the aircraft forward-based at RAF Mildenhall, Suffolk, in the UK. (*Take Off*)

More than 100 aircraft of some seventy different types appeared, including:

Aerospatiale SA.360 Dauphin
Aerospatiale Sn.601 Corvette
Aerospatiale Rallye 220GT
Aerospatiale Sport 100
Aerospatiale Tourism 100
Aerospatiale 180GT Agusta A.109
Airbus Industrie A300
BAC Concorde SST
BAC 1-11
BAC 167 Strikemaster
BAC Jet Provost T.5
Beechcraft Super King Air 200
Beechcraft King Air E90

Beech Baron 58
Bell 206B JetRanger
Boeing-Vertol CH-47C Chinook
Britten-Norman Islander BN2A-21
Britten-Norman BN Trislander 3-1
Britten-Norman Defender
Cessna Citation
Dassault-Breguet-Dornier Alpha Jet
Dornier Skyservant
Enstrom F-28A
Fokker F.27 Friendship 400
Fokker F.28 Fellowship
Gates Learjet 25B
Two Hawker Siddeley Harrier GR.1
HS 1182 Hawk
HS Buccaneer S.1
HS Nimrod MR 1
HS 125-400
HS 748
ICA-Brasov IS-29D—a Romanian glider exhibited by Singer Products
Lockheed L1011 TriStar
Lockheed SR 71 Blackbird
Lockheed C-5 Galaxy
Lockheed C-141 Starlifter
Aermacchi MB.326K
McDonnell-Douglas F-15 Eagle
Messerschmitt-Bolkow-Blohm BO105
MBB BO105C
Northrop F-5E Tiger II
Panavia Tornado MRCA P.01
Partenavia P.68
Piper Navjo
Piper Chieftain Piper Seneca Piper Cheyenne
PZL 104 Wilga 35
PZL Cobra
PZL Jantar
PZL Pirat
Rockwell Commander 112A
Rockwell Shrike Commander 500S
Rockwell Commander 685
Rockwell Turbo Commander 690A
Saab Viggen JA37

Saab Viggen AJ37
Saab 105G
Saab MFI 17
Scottish Aviation Bulldog 120
Scottish Aviation Jetstream 200
Sepecat Jaguar GR.1
Sepecat Jaguar T.2
Shorts SD3-30
Shorts Skyvan
Sikorsky H-53
VFW 614
Westland Sea King
Westland Commando
Westland Lynx
Westland Gazelle

It is of interest at this point to review Britain's aviation industry in 1974. Two important factors closely linked to the industry at this time were the escalating cost of fuel and the continued criticism—mostly from the British—of the Concorde SST project.

By 1974, the Anglo-French Concorde was coming under constant attack by certain eminent vitriolic members of society, not necessarily associated with or in the aviation field, who vehemently opposed the continuation of the project. These individuals were actively promoting the 'Stop Concorde' campaign. Of course, more than three decades later, the beautiful, sleek airliner disappeared from use with Air France and British Airways amidst waves of emotion and protestation. Even Richard Branson of Virgin Airways was prepared to keep one flying, providing he could purchase it for £1—the original cost to BOAC. Mr Branson was armed with the knowledge that in the latter years of its career, both Air France and British Airways had made handsome profits from their Concorde fleets by flying special charters at greatly inflated prices for the world's wealthy, to whom the cost of flying Concorde was completely irrelevant. Still, the cost of keeping a Concorde flying would have been undoubtedly phenomenal.

In the early 1970s, the feelings of those not directly associated with the project were much different. It was felt that members from the 'left', along with their Russian comrades, might be expected to oppose Concorde to clear the way for their Tupolev Tu-144 Charger copy. The wing-design was 'nicked' from Farnborough, although fortunately they got the wrong drawings, necessitating the fitment of the forward canards to provide stability—canards not present on Concorde, of course. The Americans too were solidly behind any attempts to slow down, ban, halt, or to scrap the Anglo-French SST. This was not unexpected, as they had decided not to proceed with their own Boeing SST design, content with their near monopoly in supplying subsonic jetliners to the free world but not wishing to lose this monopoly.

What were the motives of those in Britain who opposed Concorde? Did they, as usual, think the money should be better spent, in their opinion, on more worthy things like new hospitals and schools? Did they think that technology was getting out of hand and must be stopped? Or did they think that it would not work and would not make money? At the time, the estimated cost to build the Concorde was around £1,000 million (to be shared with France, of course). In reality, none of these arguments were justified. Already £600 million had been saved by not developing and deploying the Blue Streak LRBM, but not a single new school or hospital was built as a result of the 'saved' missile money. These sort of arguments have always surrounded such projects.

In the end, common sense prevailed; after much deliberation, in July 1974, the British government (with an annual expenditure of £40 million since the project began in November 1962) agreed to permit the completion of sixteen Concorde—eight for Air France and eight for British Airways. There were a couple of 'facts' that influenced the Labour government to continue the project. Firstly, the government knew that it would cost as much—if not more—to cancel the aircraft than to complete it, according to information issued by the Air Ministry itself and verified by an independent survey. Aircraft Nos 1 and 2 were flying, 3 and 4 were completed, 5 to 14 were up to major sub-assembly stage, and many components for 15 and 16 were already to hand. It was not a case of cancelling at this stage, but more a case of scrapping. Secondly, there was a desire to avoid a major political row with France, who, in contrast, seemed to have great faith in the project. This row would undoubtedly have erupted from outright cancellation or scrapping. Thirdly, there was a desire not to antagonise the unions involved at both Filton and Weybridge, especially at a time when Labour needed all the allies it could get.

From a technological point of view, it was important that Concorde continued in order to provide the know-how to build the less conventional airliners that would undoubtedly need to be built to replace the more conventional types that would continue in use to around 2050. By 2006, Boeing's 787 Dreamliner and flying-wing concept—the latter designed at the Cranfield College of Aeronautics, Bedfordshire, in the UK—led the way.

Another subject resulting in much debate at the time was the proposed rationalisation of the aircraft manufacturing companies in Britain. As it was mainly a political whim of the Wilson Labour government to be 'pushed' through by his Trade and Industry Secretary, Mr Anthony Wedgewood Benn, and the decision essentially reversed with the later formation of British Aerospace, it has no bearing on this review of the SBAC. However, it is worth noting that Mr Benn was a strong advocate of continuing Concorde—it was he who insisted the 'e' be added to signify the joint cooperation, strangely to much French indignation. For the record, it has to be acknowledged that many of the great post-war British achievements in aviation have happened during Labour governments, not Conservative or coalition administrations.

A more significant event in August 1973 was the launching of Airbus Industrie A300B4 and the new Hawker Siddeley HS 146 regional feeder liner. The HS 146, Britain's first major civil project for ten years, was the culmination of fourteen years of paper projects, designs, brochures, sales tours, and a search for a suitable engine by de Havilland staff at Hatfield. Hawker Siddeley had inherited the design when de Havilland was absorbed into Hawker Siddeley Aviation (HSA) under the rationalisation programme.

Following the official go-ahead, work on the HS 146 proceeded apace at Hatfield, the major design centre, at HSA Manchester, who designed the wing trailing-edge structure, the moving surfaces, and the complete tail unit, and at HSA Brough, who designed the nose-fuselage structure. The machining of the first components started in April 1974, in accordance with the production programme. Negotiation with a number of sub-contractors resulted in Shorts making the engine pods and Aerospatiale making sections of the wings, with all to be delivered to Hatfield for final assembly, where a new assembly hall had been built but never used. For many years it housed a -146 military transport demonstrator that had failed to secure any orders or much interest.

The first A300 European Airbus had entered service with Air France in May 1974 on the London–Paris route. During its first month of operation, it carried over 20,000 passengers at an average load factor of 79.3 per cent. The developed A300B consistently performed above its initial guarantees.

Due to be displayed for the first time at Farnborough in 1974 was the Shorts SD3-30 commuter airliner, the passenger-transport potential of the light transport Shorts Skyvan being realised in the Skyliner version. Originally, Shorts intended

Airbus Industries at SBAC Farnborough. The order for its initial series A300B4 gave the company the impetus it needed to mount a challenge to Boeing's decades of dominance since the launch of its first passenger jet in the 1950s, the 'Dash 80' Boeing 707. (*Aviation News*)

to develop its twenty-two-seat Skyvan into a twenty-six-seater. However, changes in the United States requirements for the commuter market dictated thirty-seat designs, and Shorts produced the SD3-30 to this capacity. Although the aircraft is basically a stretched Skyvan, with the same fuselage cross-section, major structural changes were made. The end result gave rise in essence to a brand new aeroplane—the only one at that time specifically developed to meet the US requirements and more than a re-engined variant of an existing design.

The Shorts SD3-30 was, in fact, an international collaborative venture, with Canadian-built engines, Fokker-VFW-built outer wings, ailerons, wing-tip fairings, outer flaps, and struts, and a BAC-designed-and-built centre section. Many system components were of American origin so that they would be readily available to US operators. Other collaborative ventures by Shorts at this time involved the production of components for the Lockheed TriStar, Boeing 747 Jumbo Jet, German VFW 614, and the Fokker F-28 Fellowship. Shorts' work in the Fellowship was particularly extensive; as a full risk-sharing partner, they designed and later manufactured the wing.

To meet their deadlines, Shorts needed to invest in the specialist facilities at its Airport Road Factory in Belfast, for the design and manufacture of jet engine pods. This resulted in the company becoming acknowledged as Europe's largest specialist in this field, and contracts were obtained for 'podding' the RB 211 engines for the Lockheed TriStar, the M.45 engines of the VFW, and the ALF 502-H engines of the BAe 146. Shorts were also active in the missile field.

At this time, with newly emerging community noise regulations being applied both in Britain and the United States, it was 'hush kits' that were to the fore.

The Boeing 747 jumbo jet dominated the civil wide-body passenger-jet market until Airbus Industries established its fly-by-wire A330 range. (*Author*)

At Hurn, Bournemouth, BAC's own development 1-11 was flying with 'hush kits' on its R-R Spey engines, while at Hatfield, Hawker Siddeley and Rolls-Royce were cooperating as to the best method of meeting the requirements on the HS 125 executive jets. Investigations included bench and flight testing of various suppression nozzles and the fitting of an acoustic blanket in the jet pipe. The various methods available were subjected to an intensive development programme to ensure that the final equipment adopted would give the noise attenuation required to comply with the regulations and minimum effect on performance. The hush kits were installed on new-build HS 125s and were available for retro-fitment to Series 600 aeroplanes already in service.

On a national scale, for many years, the MoD (in conjunction with the Department of Trade and Industry) conducted an expanding research programme into problems associated with the reduction of aircraft noise.

For part of this programme, a specially-equipped HS 125 was used to study the effectiveness of airframe shielding against engine noise in flight. For this purpose, the 125 executive jet—basically a standard Series 1 aeroplane—was fitted with special Hartmann noise generators that could be sited in various positions on the airframe. Also contributing to the noise reduction programme was the Handley Page HP115 experimental aircraft, which was similarly equipped with noise generators, and the Miles Student. The Student, after a storage period of eight years, was commissioned to assist in the programme, investigating noise suppression by airframe shielding. The refurbishment and modification of the Student was undertaken by Miles Dufon. The recommissioned aircraft flew for the first time on 9 January 1974, to be operated by Hawker Siddeley under a MoD contract.

A welcome contribution to the battle against noise at this time was the Rolls-Royce Spey 67C. A development of the Spey Mk 512, the first 'quiet' Spey was delivered in late 1976.

By 1974, a number of new military types had entered service. The first RAF Sepecat Jaguar was delivered in May 1973 to RAF Lossiemouth, Scotland, for ground-crew training. This was followed in September 1974 by further aircraft for flying training with the Jaguar OCU and the formation of the first Jaguar squadron (No. 54(F)) at Lossiemouth. Thus began the build-up that made the Jaguar the biggest single element of the RAF's front-line strength until another collaborative venture type, the Panavia MRCA, entered service. The RAF received 200 Jaguars—165 GR.1 single-seat strike planes and thirty-five T.2 two-seat trainers—and the type equipped nine squadrons, five of which were based in Germany for NATO and four of which were home-based. The Sepecat Jaguar also became operational with the Armée de 'l Air in June 1973.

The reaction of pilots to the fighter was extremely encouraging. Sir Denis Smallwood, AOC-in-C of RAF Strike Command, remarked:

The Boscombe Down DRA experimental Sepecat Jaguar T.2 Raspberry Ripple. (*Author*)

> Performance in the air is highly satisfactory, particularly in terms of the great weapon load—10,000 lb more than early marks of the Canberra which this aircraft can carry, in its conventional capability, in its range of 1,800 miles, and not least, perhaps in its fine airfield performance

The introduction of the Jaguar brought a significant increase in the RAF's total ground-attack capability. The Jaguar was specially designed for low-level operations, and this, coupled with its exceptional range, made it ideal for modern long-range reconnaissance made almost exclusively at low level. BAC produced a special pod for the aircraft that could accommodate a wide variety of optical and infra-red cameras.

Rolled out in early 1974 was the first prototype of the swing-wing Panavia MRCA, which was destined to become the key front-line aircraft of the RAF and the air forces of Germany and Italy for the remainder of the twentieth century and, in its upgraded form, well into the new millennium.

The two-seat Tornado (developed under a tri-nation agreement between Britain, Germany, and Italy) was described as a 'multi' aeroplane to fulfil six major roles—close air support, interdiction, strike, land-based naval strike, air superiority, and interception and reconnaissance. In RAF service, the Tornado was intended to replace the Avro Vulcan, the HS Buccaneer, and the MDC F-4 Phantom II.

With a planned initial production of about 800 aeroplanes, one area of concern was the unit cost of the aeroplane—£3.5 million, which included research and development. However, in terms of modern combat aircraft, when compared with the price of the Grumman F-14 Tomcat for the US Navy (over £8 million), the MRCA was a bargain.

The first prototype Panavia MRCA, P.01 (German-built and civil-registered as D-9591), flew for the first time at Manching on 14 August 1974, under the control of BAC chief test pilot Paul Millet; the rear seat was occupied by MBB's Nils Meister. The second prototype, P.02 (UK military serial XX946), flew from Warton, in the UK, two months later, followed by seven more prototypes spread across Britain, Germany, and Italy. These nine aeroplanes were followed by six pre-series aircraft flown by service pilots at government test establishments in the three countries. The first production order (forty aircraft) was authorised in July 1976. The aircraft were then ordered by NAMMA—the NATO MRCA Management Agency—in batches of about 160 aircraft at approximately eighteen-to-twenty-four-month intervals, with each batch containing a proportion of aircraft for each of the three original customer nations. Deliveries began in 1980, and the last of 795 IDS/ECR variants (including seventy-two for Saudi Arabia) were delivered in 1992. In addition, 197 Air Defence Variants (ADV) variants were built, including twenty-four for Saudi Arabia.

In June 1967, the six-day Israeli-Arab war showed for the first time the importance and terrifying efficiency of the latest generation of missiles. On the ground, Israeli armour suffered grievous losses to the Russian-supplied Snapper and Sagger anti-tank missiles.

British contenders in this field included the BAC Vigilant and the more advanced long-range Swingfire, which entered service with the Royal Armoured Corps in 1970. There was also an airborne version, known as the 'Airstrike Swingfire', under development. It was the fixed-site ram-jet Soviet SA-6 surface-to-air missiles that dominated over the Suez Canal, but in Sinai it was the infantry-launched SA-7 surface-to-air SAM weapon. Britain had no equivalent to the former, but in the latter there was a counterpart—the Short Blowpipe, a supersonic weapon that weighed less than 50 lb, meaning it could be carried and fired by one man. It was intended primarily for the defence of forward troops against low-flying aircraft, but it was also capable of dealing with lightly armoured vehicles and small surface vessels. In the man-portable role, the operator supports the launch tube on his shoulder, sights the target through an integral monocular sight, and steers the weapon by using a thumb stick.

Short Seacat and Tigercat short-range surface-to-air missiles were also in production for the Royal Navy. The Seacat was the world's most widely used ship-borne missile system, being in service with the RN and fourteen other nations. The Tigercat was the land-based version of Seacat and could be operated with a wide range of control units for specific applications; the radar-enhanced variant incorporated a Marconi ST 850 target-tracking radar and television to provide dark and blind-fire capabilities.

The Seawolf was a Seacat development by BAC, to equip Royal Navy Type 21 and Type 22 frigates. This weapon was intended for point defence against anti-ship missiles and aircraft in poor weather, rough seas, and in close proximity to coastlines. Test rounds were first fired in 1973, though escalating costs meant that the programme was in jeopardy at this time.

Meanwhile, the production of the first BAC (BAe) Rapier mobile surface-to-air missile continued; the system was already in service with the British Army and RAF Regiment for airfield defence, being noted for its quick reaction time. The Rapier is still used (in upgraded form) today. The basic system comprises two Land Rover-towed trailers—one carrying the launcher, tracker, radio, and four missiles in containers, and the other carrying nine more missiles and additional equipment. The missile can be fired by one man, with another in attendance to assist in reloading.

A new air-to-air missile, the XJ 521, was under development by Hawker Siddeley Dynamics (HSD). Officially announced on 3 March 1974, the all-weather weapon was based on the US Raytheon AIM-7 Sparrow but offered substantial improvements—particularly in range and performance. Initially, it was intended that the missile would replace the Ferranti Red Top carried by RAF Lightning air defence fighters, later to equip MDC F-4 Phantoms and the MRCA. However, this did not materialise; the Phantom and ADV Tornado carried the Raytheon weapon instead.

Space-related activities in the UK were overshadowed by the concluding Apollo 17 mission to the moon, the launching of NASA's Skylab in 1972 (with its subsequent manning by three crews), and by Russia's Salyut 3-Soyuz 14 mission. Nevertheless, Britain had not been inactive during this period; the most significant achievement was the securing by HSD of a substantial £5 million contract for part of the space lab to be carried into orbit by the space shuttle under development in the United States. The overall £95 million contract went to the nine-nation consortium led by the German company Erno.

Another British space venture was the successful launch of the UK-X4 experimental satellite on 9 March 1974 by a US Scout rocket. Built by HSD and Marconi, the UK-X4 was a three-axis stabilised spacecraft intended to demonstrate the technology involved in the design and manufacture of this type of experimental platform for use on small spacecraft. The UK-X4 was the eighth British satellite to be launched by NASA and the fourth by a Scout vehicle. The spacecraft was controlled by the Satellite Control Centre, RAE, with assistance from NASA's Space Tracking and Data Acquisition Network.

At this point in time, it was generally considered that the British aerospace industry was in a position from which 'technically it could face the future with genuine confidence'. This was a significant turnaround when compared to 1969, just a few years before, when the then-President of the SBAC was quoted as saying:

> There can be no doubt that somewhere in high places there is, and has been for some time, a desire to confine, if not eliminate, this industry.

From 1974 onwards, SBAC Farnborough continued to expand and became the 'shop window' for all the world's leading aerospace manufacturers, with international participation from both east and west. However, it was not until 1986 that the Russians participated, when the 392-ton Antonov An-124 Condor dominated the static park. In the 1970s, it was still only Britain, the USA, and Russia that possessed 'total capability'—the ability not only to design and make aeroplanes, but also all the equipment that went into them. This allowed Britain to build many of the types specifically required by the RAF, saving money on imports while simultaneously offering the opportunity of export earnings—not only of complete aircraft, but of engines and equipment for foreign airframe manufacturers to put in their machines. Many air arms and airlines operated aircraft powered by Rolls-Royce engines. For instance, every time Fokker sold an F-27 Friendship, 40 per cent of the money came to Britain. The type was powered by R-R Dart turboprop engines equipped with Rotol propellers, Dowty landing gear, and other British equipment.

For the aviation buffs, the shows of the late 1950s and early 1960s were the most memorable, as the relaxed flight-safety regulations saw more 'seat-of-the-pants' flying by modern-day standards; the pilots virtually had complete freedom with the execution of their display flying. It was also a period of great innovation, with a never-ending array of weird and wonderful designs and prototypes, especially for the military, as the Cold War took grip. For the first time, there were rocket motors and boosters, afterburners, and high-speed military research

The massive Antonov An-225 Myra, a direct development of the An-124 Condor. Only one was built. It is said that the Myra could accommodate all the Red Arrows display aircraft at the same time in its massive cargo hold. (*Author*)

planes such as the Fairey Delta 2 breaking the sound barrier (although sonic booms were banned in 1955).

Farnborough has witnessed a number of accidents over the years; some were relatively trivial incidents, such as a wheels-up landing, while others were more serious, such as the catastrophic break-up of the prototype DH 110 (serial WG236) flown by John Derry on 6 September 1952. The prototype crashed, killing Derry, his flight observer, Tony Richards, and twenty-eight members of the public. The other prototype DH 110 was grounded pending investigation, modifications, and other design work to enhance the aircraft's flight characteristics, handling, and naval specification requirements. When production was switched from Hatfield to de Havilland's (formerly Airspeed) Christchurch plant, the work was carried out under the auspices of W. A. Tamblin. A third prototype DH 110 was named 'Sea Vixen', the type entering service with the Royal Navy in June 1957. After a number of tragic accidents at other displays, it was not long before strict new rules were introduced regarding display flying at air shows. The focus was on parameters such as minimum height and the separation between the display line and the crowd, with no display flying to be made directly over the crowd's heads. These rules and guidelines have been strictly enforced ever since.

As the new regulations began to limit the amount of risqué flying and with the enforced gradual decline of new types on show, the SBAC began to lose popularity until 1988, when the Russians first appeared with their Mikoyan MiG-29 Fulcrum and Sukhoi Su-27 Flanker fighters. The former demonstrated its rapid climb to altitude and subsequent tail-slide, while the latter demonstrated its breath-taking 'Cobra'. Both of these highly manoeuvrable fighters have continued to thrill the crowds ever since—not only at the SBAC, but at Paris, Berlin, and the Fairford International Air Tattoo (IAT). However, at Fairford, the Mikoyan MiG-29 suffered an ignominious (but luckily non-fatal) collision.

During the same period, probably the most awe-inspiring machine from the USA was the Boeing B-2A Spirit flying-wing stealth bomber. From the European industry, it was the Saab JAS 39 Gripen, the French Rafale, and the Eurofighter Typhoon—all delta-wing designs with forward canards for enhanced stability.

Farnborough International '80, the thirty-second SBAC, was held from 1–7 September and was one of the most successful shows. It fortuitously enjoyed one of the few patches of good weather afforded to Britain that year. The orders for Britain's aerospace industry announced during the Farnborough week totalled £286,500,000—compared with a little over £50 million for the 1978 show. A significant contract was awarded for the NADGE (NATO Air Defence Ground Environment) radar network, in which Marconi and Plessey were heavily involved. According to the SBAC, potential business or deals clinched at Farnborough were expected to peak at a value of around £500 million. Attendance was up too, totalling over 217,000 for the three public days alone. So too was the number of entries, which rose to around 150 aircraft, with a wide representation from abroad; the Americans returned in force, although China and Russia still played no part at this time.

Some 'old-timers' suggested that the flying lacked some of the 'zip' of earlier times, when a great number of service demonstrations were included. However, this was probably nostalgia, as in general, displays were very sensibly designed to illustrate the true roles of the aeroplanes involved. Trainers showed the training manoeuvres; helicopters displayed their unique capabilities; and STOL was to the fore, not only in its military applications (with Hawker's chief test pilot, John Farley, undertaking a steep VTO in the BAe FRS.1 Sea Harrier) but also, where relevant, in the civil field, for airliners large and small (a matter would continue to be discussed into the twenty-first century). Quietness was the relevant factor emphasised by large and small airliners operating in and out of sensitive locations, where compliance to strict noise regulations was being applied.

With fuel economy being of increasing importance for operators, there was a resurgence of turboprops on display in the air. As usual, smoke was used to enhance aerobatic displays; particularly interesting from this aspect was the General Dynamics F-16C Fighting Falcon, whose wing-tip smoke generators really showed the surprising agility and manoeuvrability of this aeroplane. The Red Arrows displayed in their new mounts, BAe Hawks, for the first time, ably supported by their own patriotic red, white, and blue smoke generators, emphasising their typically impeccable display. Meanwhile, three Royal Navy Sea Harrier FRS 1s made a 'big bang' in a dashing simulated ground-attack routine.

Among the nearly fifty types making their debut at Farnborough '80 was the Panavia Tornado F.2A ADV, forerunner to the series production F.3, with its highly capable AI.24 Foxhunter GEC/Marconi multi-mode radar. Also present was the other, less successful type, the HS Nimrod AEW.3, also intended to be equipped with a GEC/Marconi radar. The Tornado F.2A ADV was accompanied in the air by the Panavia Tornado GR.1 IDS version, which was well-laden with stores. The GR.1, entered service with the Tri-national Tornado Training Establishment (TTTE) at RAF Cottesmore, Rutland, in July 1980. It eventually equipped Luftwaffe and Marineflieger units in Germany, the AMI and MMI in Italy, and the Royal Saudi AF strike units (complementing their MDC F-15 Eagles). At Farnborough, the GR.1 demonstrated its great manoeuvrability in short bursts.

Highly significant combat types from abroad included Dassault's Mirage 2000 and the even more impressive Super Mirage 4000—the latter never entering series production, not even for the Armée de 'l Air. Also present was the McDonnell-Douglas F-15E Strike Eagle, an F-15B two-seater developed at company expense to provide a potent all-weather/night-strike multi-role aircraft. It had flown non-stop across the Atlantic from Maine to RAF Mildenhall, Suffolk. The same company's dual-seat TF-18A Hornet for the US Marine Corps was also present at Farnborough, in addition to the Saab Viggen in its JA37 fighter-optimised version, emphasising its great manoeuvrability and STOL qualities.

Due to its diminutive size, the Saab JAS 39 Gripen is stealthy by default. (*Author*)

Above: A No. 800 NAS AIM-9 Sidewinder-equipped BAe Sea Harrier FRS.1 *en route* to the South Atlantic in 1982. (*Author*)

Below: Fifth-generation canard-equipped Dassault Rafale land-based and carrier-borne interceptors equip the *Armée de l'Air* and *Aéronavale* in the new millennium. (*Author*)

The original BAe Hawk T1 trainer equipped seventeen foreign air forces in addition to the RAF. (*Author*)

Above: The fighter-optimised Saab JA37 Viggen in European 'low-vis' grey. (*Author*)

Below: Dassault Mirage 2000-series aeroplanes replaced the similar delta-winged Mirage III in *Armée de l'Air* front-line service. (*Author*)

The Boeing/MDC F-15E Strike Eagle is a dual-seat development of the MDC F-15B Eagle. (*Author*)

The Scottish Aviation Jetstream 31 gained a foothold in the USA for short-range regional flights; the type is still operational with a number of US carriers in the new millennium. (*Warplanes collection*)

Aermacchi displayed its new MB.339K trainer, which was named 'Veltro 2' in honour of the Macchi MC.205 Veltro ('Greyhound') fighter of the Second World War.

On the civil side, apart from the appearance on some days of a British Airways Lockheed TriStar L1011-500, the show was dominated by an A300 Airbus in Singapore Airlines livery.

Small turboprop airliners appeared in great profusion: Shorts obtained several new orders for these airliners during the show; BAe showed its Jetstream 31; Pilatus Britten-Norman showed their new turbine Islander; and there were two Dornier 128 versions and a Dornier TNT. In addition, there were four small turboprop airliners from Embraer—two Bandeirantes and two Xingus—as well as an EMB-111 Maritime Bandeirante, Embraer starting to make inroads into the UK market at this time. The once-unknown Brazilian firm had produced 2,211 aeroplanes by 30 June 1980. STOL experts de Havilland Canada displayed their Twin Otter, Buffalo, and Dash 7. Canadair's new Challenger, designed for the executive jet market, opened the flying display on several days. DH Canada's new Dash 8 was one of several new small transport planes displayed in model form in the tents. While the show was on, Saab announced it had received the first order for its new thirty-four-passenger commuter airliner, the 340. Switzerland's Crossair had ordered five of the type plus five options; Stillwell Aviation of Australia followed by ordering twelve, and Swedair signed a letter of intent for seven.

On the rotary-winged front, Westland provided their traditional 'circus' in the air, receiving an order for twenty-four more British Army Lynx AH 1s. Westland also generated significant interest from British Airways Helicopters in its new WG.30 and Sea King replacement the EHI EH 101, which would be developed by the company in conjunction with Agusta of Italy.

A DHC Dash 6 Twin Otter light transport of the Spanish Air Force. (*Author*)

The two-to-three-crew DHC Dash 7 CC-132 utility transport was operated by the Canadian Armed Forces in Europe as communications aircraft until the withdrawal of the CAF from West Germany.

The DHC Dash 8M Boeing Canada E-9As were used as Airborne Telemetry platforms and as missile-range-control aircraft under civilian contract to the USAF. (*Author*)

Seven two-crew Canadair CL601 Challengers were procured by the German Luftwaffe; some were VIP/VVIP configured. (*Warplanes collection*)

The Luftwaffe also operated Dornier 228 light transports—not in a military role, but for environmental reconnaissance. (*Luftwaffe*)

11

Farnborough '80: The Best International Yet

Farnborough '80 was probably the best international Farnborough yet, with the main new types on display including:

Aerospatiale AS.355E
Aerospatiale SA.365N Dauphin
Aerospatiale SA.341G Gazelle
Aermacchi MB-339K Veltro 2 (company development of the MB-339A)
Airbus A300B4-203
Agusta A.109
BAC Strikemaster Mk 80A (demonstrator)
BAe Bulldog 124 (company demonstrator)
BAe Jetstream 31 (prototype conversion)
BAe Sea Harrier FRS.1
BAe Hawk T.1
BAe (HS) Nimrod AEW.3 (serial XZ286, production prototype)
Bell 249 YAH-1S (a development Huey Cobra 'gunship')
Beech Super King Air 200
Boeing E-3A-BN Sentry AWACS (the fifth production aircraft)
Dassault-Breguet/Dornier Alpha Jet
Dassault Mirage F.1C (French AF EC 1/5)
Dassault Mirage 2000 (×2)
Dassault Super Mirage 4000
Dassault Falcon 50
DHC CC-132 Dash 7 (the second production aircraft for the Canadian Armed Forces)
Fairchild A-10A-FA Thunderbolt II (USAFE, serial 76-126)
Fokker F-27M Maritime Friendship (a new aircraft for the Angolan AF)
Fokker F-28 Fellowship (a production aircraft for Garuda Indonesia Airlines)

Gates Learjet 35A (a maritime multi-mission prototype with radar under forward fuselage)
General Dynamics F-16B Fighting Falcon
HS 125 Series 700B
HS 748 Series 2A
Lockheed L1011-385 TriStar 500 (a production aircraft in service with British Airways)
MBB BO 105CB
MDC TF-18A Hornet
MDC F-15B-4-MC 'Strike Eagle' (the second prototype)
Panavia Tornado F.2 (ADV version, the first prototype)
Panavia Tornado GR.1 (RAF, serial XX947)
PBN BN-2T Islander (a development aircraft)
Saab JA37 Viggen (Swedish AF)
Sepecat Jaguar International (an RAF aircraft, serial XX979, retained at Warton for development use)
Westland Sea King HAR 3 (RAF, the third production aircraft, serial XZ587. Used by Westland Helicopters Ltd for development)
Westland Sea King HC 4 Commando (Royal Navy, serial ZA297)
Westland AH.1 Lynx (company demonstrator, serial ZB500)
Westland HAS.2 Lynx (the fifty-first series production aircraft for the Royal Navy, serial (XZ727)
Westland WG-30 (the first prototype, registration G-BGHF)

By the 1980s, Farnborough had become more than just a shop window for new aircraft; it was also a great technical show, with a wealth of new technology to be found amongst the static displays of equipment, models, videos, simulations, and so on.

In terms of flying, Farnborough '86 was probably the vintage year. The company-funded BAe EAP technology demonstrator (forerunner to the Eurofighter Typhoon) and the French Dassault Rafale were displayed, but of greater significance was the world's first fly-by-wire airliner, the Airbus A300 demonstrator (civil registration F-BUAD). Under the command of Gordon Corps, the demonstrator very graphically showed what a large aircraft could do when equipped with a computer-controlled flight system; it hung on to the back of the drag curve in an almost unbelievable pitch angle. In the tents, the Airbus family was displayed in model form, including the A300-600, A310, A320, A330, and A340.

Among the many other interesting types displayed was the mock-up of the AgustaWestland EH 101 multi-role helicopter. Unfortunately, due to the tragic loss of the sole prototype BAe Hawk 200 single-seat fighter on 2 July 1986, the mock-up was displayed.

The British government had committed to the European Fighter Aircraft programme with Spain, Germany, and Italy, and so the relevance of the BAe EAP

For a fighter aircraft, the Eurofighter Typhoon is a 'big delta', but in its Tranche 3 variant it is now a superb multirole combat aircraft. It is deployed to RAF Mount Pleasant to defend the Falkland Islands. (*Air Power International*)

A two-seat combat-capable Dassault Mirage 2000B. Due to the Rafale's inordinately long planned gestation period (mainly due to lack of government finance), the Mirage 2000 continued in front-line service well into the new millennium. (*Warplanes collection*)

became apparent when the aeroplane was displayed with a complete weapons fit similar to the one proposed for the EFA.

Dassault-Breguet's Rafale, an experimental demonstrator, made an impressive display, with production aircraft intended to replace Armée de 'l Air Mirage IIIEs, Sepecat Jaguars, *Aéronavale* Vought Crusaders, and Dassault Etendard IVPs.

A first-time visitor to the show was the new forty-eight-to-fifty-eight-seat Fokker 50, a follow- on development to the F27 Friendship. In the civil field, another 'first' was the BAe-146QT (Quiet Trader), which was able to accept standard air cargo pallets through its large starboard fuselage freight door.

In July 1986, British Aerospace revealed plans to convert its BAe 146 Series 100 prototype into an aerodynamic and structural prototype of the Series 300, the lengthened 100-passenger version scheduled to fly in 1987. A model of the Series 300 was displayed at Farnborough alongside another very successful type from the BAe stable—the BAe 125 Series 800 executive jet, which first flew in May 1983. Also of note is that by the end of August 1986, sixty-three firm orders had been secured for the new French high-winged, fifty-seat ATR 42 twin-engined turboprop regional transport.

Although it did not fly during the show, the Soviet Antonov An-124 Condor landed on British soil for the first time on the morning of Saturday 30 August. Further British firsts included the BAe ATP sixty-to-seventy-two-passenger regional, which had only flown for the first time on 6 August, and the new Chichester-Miles

Dassault's Rafale demonstrator was designed at their St Cloud office, Paris. It was slightly larger than the series aircraft now in service with the *Armée de l' Air and Aéronavale*. (*Warplanes*)

The Dassault-Breguet Etendard IVP differed from the IVM version in being a carrier-borne reconnaissance aeroplane, with under-fuselage cameras replacing the earlier type's combat electronics and cannon. (*Warplanes collection*)

Two BAe 146-100 evaluation aircraft were operated by No. 10 Squadron at RAF Benson, Oxfordshire for two years prior to the purchase of three series aeroplanes for the Queen's Flight. (*Author*)

Consultants (CMC) Leopard, an unusual four-seat business aircraft of mostly GRP composites construction and powered by two small Noel Penny Turbines NPT 301-3 turbojets.

Military firsts included the first visit of the Swiss Pilatus PC 9 turboprop trainer (registration HB- HPA) to Farnborough and the Aeritalia/Aermacchi/Embraer AMX strike-attack fighter, which entered series production in July 1986. Another Pilatus type was the Britten-Norman AEW Defender (registration G-TEMI), with its large nose radome housing Thorn EMI Skymaster radar. The Defender was not successful in securing any orders for the company, although it was still on offer more than two decades later, with large Westinghouse radar in a different housing.

Farnborough '88 was the fortieth birthday of the show. By now, in addition to its flying and static aircraft displays, the show had no less than four exhibition halls with 54,000 square metres of floor space.

This year marked the first time that the ubiquitous Soviet Mikoyan MiG-29S Fulcrum air defence fighter and its new-technology 'glass cockpit' MiG-29M counterpart had appeared outside the Soviet Union; both aircraft put on a superb flying display. Also from the Soviet Union, on its second visit, was the Antonov An-124 Condor; unfortunately, it had to abort its flying display due to an engine problem that kept the type grounded throughout the show, with a new engine having to be flown in from Russia. However, it was subsequently 'opened' to the public in the static display.

Above: The Pilatus PC 9 was a direct descendant of the similarly shaped and proven PC 7. (*Author*)

Opposite middle and below: The Antonov An-72P Coaler was originally conceived as a medium transport in competition with the Airbus FLA (A400M), but with its high-set jets, it has been adopted by the Russians for maritime-patrol duties. (*Aviation News*)

Above: The Embraer EMB AMX-1 is colloquially known as the 'Pocket Tornado'. In upgraded form, the type remains in service with the AMI and Brazilian Air Force in the new millennium. (*Author*)

12

The Year of the Russians

Farnborough '92, which opened on 6 September, was 'the year of the Russians'. The newly formed Commonwealth of Independent States (CIS) arrived with twenty-one aeroplanes and a technical team that included some of the leading designers in the industry, from the countries' major airframe and engine organisations. All had an overwhelming desire to cooperate with Western companies in the future development of commercial aviation products. The first evidence of this was in the Tupolev Tu-204 medium-haul airliner, which made its maiden flight on 2 January 1989 with twin under-wing Aviadvigatel PS-90A turbofan engines. On 14 August 1992 it had flown for the first time after being fitted with two Rolls-Royce RB211-535 engines. The Bravia RB-211-powered aeroplane (registration CCCP-64006 but with the new CIS markings on the tail) flew daily at the show.

For Russian aircraft aviation buffs, Farnborough '92 was the place to be. The types on show included: the Antonov An-124 Condor; the An-72P Coaler; the Ilyushin Il-76 Midas; the Kamov Ka 50 Hokum (Black Shark) attack helicopter; the Mil Mi-17M Hip; the Mi-26 and Mi-34 Hind; the Mikoyan MiG-29S and MiG-29M Fulcrum; the MiG 31 Foxhound; the Sukhoi Su-24MR Fencer; the Su-25T Frogfoot; the Su-27UB Flanker; the Su-29 and 29UT aerobatic trainers; the Su-35 Flanker; the Tupolev Tu-22M-3 Backfire; the Tu-204; the Yakovlev Yak-38 Forger; the Yak-112; and the Yak-141 Freestyle.

Dominating all other aircraft in the static display was the Russian Naval Aviation 124-tonne Tupolev Tu-22M-3 Backfire C medium-range strategic-bomber. The bomber gave only one flying display during its visit to Farnborough, during which it made a high-speed pass with its variable-geometry wings in the fully-swept position. This was followed by a slower pass, with the wings fully spread 20 degrees to their full 126-feet-6.5-inch span.

It probably helped détente when the UK MoD advised the designer of the Kamov Ka 50 attack helicopter, Sergei Mikheyev, that the helicopter would be seriously considered to meet the UK's requirement for such a machine. However, by 2006, only

The Kazan Mi-17M Hip, displayed here in Russian Navy (AV-MF) livery, is a radar-equipped all-weather ASW helicopter. (*Author*)

The Sukhoi Su-25T Frogfoot is a highly capable ground-attack aircraft that was combat tested and evaluated in the Soviet war in Afghanistan in the 1980s. (*Warplanes collection*)

Above: A Sukhoi Su-29 aerobatic aircraft sporting a Soviet 'Red Star'. (*Author*)

Below: The Sukhoi Su-27 Flanker 'Russian Knights' display team visited RAF Scampton in Lincolnshire, home of the RAF Red Arrows. (*Air Power International*)

Above: A Sukhoi Su-22 fitter of the 'Cold War' Czechoslovakian Air Force. (*Author*)

Below: A Tupolev Tu-22M-3 Backfire of the Confederation of Independent States (CIS) AV MF. (*Air Power International*)

The Mil Mi 34 Hind attack helicopter is the export version of the Russian AF (formerly Army Aviation) Mi-24. (*Author*)

Above and below: The 'glass cockpit' Mikoyan MiG-29M Fulcrum, the backbone of the Russian Federation Air Defence Force along with the dual-seat MiG-29UB, seen here in Hungarian AF livery before being usurped by the leased Tranche 3 Saab JAS 39 Gripen. (*Author*)

a handful of the original evaluation machines were in use with the Russian Federation Air Force—the aircraft had been usurped by the new Mil Mi-28 Nighthawk. The Hokum arrived at Farnborough in an Antonov An 124 Condor, and following assembly it was parked and remained firmly ground-based throughout the show, apparently due to a dispute between the Kamov personnel and UK immigration. Made of over 35 per cent composites, the Ka 50 fuselage is one of the most densely engineered rotary-winged designs ever. Access panels covered the surface, and its single 30–mm 2A42 high-velocity cannon was mounted in a hydraulically operated barbette on the lower centre fuselage. The single-seat helicopter, with its fighter-like cockpit, was fitted with a head-up display and single stick and, for the first time for a rotary-winged aircraft, a unique crew-extraction system. The escape sequence involved severing all six rotor blades and jettisoning the cockpit roof panels before a large single rocket unit fired the pilot clear 0.3 seconds later, with seat separation and automatic parachute deployment within 2.3 seconds.

The newly-styled Yak Aircraft Corporation of Russia brought its Yakovlev Yak 141 V/STOL Freestyle, which made its first flight and world public debut on Thursday afternoon, with Vladimir Yakimov at the controls, to avoid the adverse weather conditions earlier in the week. The 141 was a very slab-sided machine that certainly conceded nothing to stealth. However, first flown in March 1989, the Freestyle did have fly-by-wire controls and was powered by a single Soyuz R-79 vectored-thrust afterburning engine. For STO/VL operations, it was augmented by two Rybinsk RD-41 lift engines. Wing-tip ejectors give control in the rolling plane, while the swivelling VTOL engines were controlled by fly-by-wire handled pitch. With the 141 came a first-generation Yak-38A Forger V/STOL aircraft of the former Soviet Navy; in reality, it was a much better aircraft than it was generally given credit for by the West. Of similar vintage to the early Sea Harrier, the Yak-38 was also flown for the first time in the West at Farnborough '92.

The two Russian fighter bureaus, Mikoyan and Sukhoi, displayed their latest MiG-29 Fulcrum and Su-27 Flanker variants. The former was represented by two variants—a MiG-29S high-altitude fighter, which was already in service with two CIS Air Force fighter regiments and reportedly had much better combat potential than the earlier operational versions, and its new technology MiG-29M with quadruplex digital fly-by-wire. At the time, the 'M' was still in the late stages of development. It was being promoted as a multi-role type with advanced radar for both air-to-air and air-to-surface detection of targets. In the cockpit, two multi-functional displays were fitted in addition to the head-up display. With an extra 1,500 litres of fuel and improved handling and agility, its zoom climb to altitude, tail slide, and recovery were extremely impressive.

Also present from MiG MAPO was the mighty MiG-31 Foxhound, which had first been shown to the West at Le Bourget, Paris, in 1991. Mikoyan chief test pilot Valery Menitsky stressed that the aircraft's Zaslon phased-array radar was effective against stealth aircraft, giving broad hints that it had been evaluated against the as-yet-unseen Russian stealth aircraft (which eventually surfaced in 2013).

Sukhoi displayed its new 'glass-cockpit' Su-35 version of its Flanker. Externally, the aircraft looked very similar to the standard Su-27, but it embodied a range of improvements that general designer Simonov declared rated 'with the best in the West'. A new fly-by-wire system and canard foreplanes gave the Su-35 improved agility in combat and the pilot's workload was improved by having three multi-function cockpit displays and an ejection seat angled at 30 degrees for high-g manoeuvres. A multi-role variant, the Su-35 incorporated ESM pods on the wing-tips and was displayed carrying the latest X-31 long-range stand-off air-to-surface missiles equipped with four ramjet boosters. Externally, the new Flanker was identified by its enlarged offset IRST housing immediately in front of the windscreen. A TIALD pod could also be carried on the engine nacelle pylon. The Super Flanker had multimode radar with both terrain-avoidance and terrain-following modes.

Also from Sukhoi came a swing-wing Su-24MR Fencer E strike bomber that was flown to Farnborough by test pilot Anatoly Ivanov, although the aircraft did not take part in the flying display. Almost identical to the reconnaissance version already in service with the Russian Air Force, this was in fact a development aircraft with an ejectable pod containing reconnaissance equipment. When recovered, the information inside the pod could be deciphered and analysed for tactical use. The aircraft carried the standard twin R-60 AA-8 Aphid missile launcher on the port-wing pylon as well as the standard infra-red and daylight reconnaissance equipment.

Other Farnborough Soviet debutantes from Sukhoi were two Su-25T Frogfoot ground-attack planes that first flew on 22 February 1975. The initial production of this aircraft took place in Tbilisi, Georgia, ending in 1989 with around fifty partially-built airframes remaining unsold.

Another Soviet type of note was the Antonov A 72P maritime-patrol variant of the Coaler. A high-lift utility twin-jet transport, the Coaler first appeared at Farnborough in 1986, when it demonstrated its remarkable agility. The An-72P is a coastal-patrol variant and was displayed with two UB-32M rocket pods on under-wing pylons, a 23-mm GSh-23L cannon in the starboard main gear fairing, and, remarkably, four 100-kg (227-lb) bombs slung inside the roof over the rear loading ramp, which had to be swung clear prior to their release. What effect this has on the aircraft's handling is not known.

In the tents, the Russians had several missiles on display. Probably the most impressive of these were the Vrimpel Amraamski RVV-AE, which was to be carried by the MiG-29M Fulcrum, and the Zvezda Kh-31P anti-radiation missile which was to be carried by the Su-35 Super Sukhoi.

In comparison, the West had little new to offer, although Saab unveiled two of its latest products by way of the JAS 39 Gripen and the fifty-eight-seat Saab 2000 twin-turboprop regional airliner—in effect, a stretched Saab 340 with greater wingspan.

Farnborough '92 represented the inaugural display for Sweden. King Carl Gustav was in attendance alongside the Swedish defence minister; the latter arrived in a Swedish Air Force dual-seat Saab JAS.37 Viggen. The minister later

The Kazan Mi-17 Hip is an upgraded Mil-8 designed for all-weather night-fighting operations. (*Author*)

Above: The Saab JAS 39 Gripen Tranche 3 aircraft that flew during the SBAC's afternoon show was the fourth prototype. The type has since gained a number of orders from foreign air forces, including the SAAF. (*Air Power International*)

Below: Saab 2000 registration SE-003, the company's third prototype, seen flying with Saab 340B registration SE-C12/N112AE in American Eagle colours. (*Air Power International*)

stated that Sweden would be prepared to sell the new JAS 39 Gripen to almost any country that had the cash to pay for it. The aircraft's flying display during its first visit to Farnborough were certainly impressing many overseas visitors, particularly in the use of its massive foreplanes to supplement the rear-side-fuselage-mounted airbrakes for short landings.

The MDC/BAe Harrier GR 7 was displayed, concentrating more on conventional flight and less on the hover than in previous years, while the Hawks and Tornados had little new to offer. Dassault displayed its Mirage 2000 and its new stablemate, the Rafale C.01, and General Dynamics displayed a Turkish Block 40 LM F-16 Fighting Falcon with fly-by-wire in lieu of the analogue flight-control system.

By 1992, the British aviation industry was already inextricably linked to Europe's. As already clearly demonstrated by the British participation in the Concorde SST, Sepecat Jaguar, Panavia Tornado, and Airbus Industrie airliner series, Britain's participation in Europe's aerospace industry could produce some of the world's best aircraft in most military and civil categories—except, perhaps, regional airliners.

In retrospect, only a handful of post-war European aircraft that were developed on a national basis have been successful, and in almost every case these were started and have disappeared decades ago. Of note were the British Vickers Viscount regional airliner and Hawker Hunter fighter, the Dutch Fokker F27 Friendship, and the French Dassault Mirage III/5 series. Of the more recent designs, the BAe Sea Harrier—as opposed to the MDC/BAe AV-8B Harrier II development—the Dassault Mirage 2000, and the Rafale have not been so successful, although Sweden's enhanced Tranche 3 JAS 39 Gripen (marketed in conjunction with BAe Systems at the turn of the century) has been faring better, with a number (though not a large amount) sold on the export market. BAe Systems was responsible for the Gripen's composite wing design.

In the never-ending struggle to remain viable and produce competitive new products in an ever-diminishing marketplace, first came nationalisation, then rationalisation, then amalgamation of most of Britain's and Europe's principal companies in the aerospace industry. Ironically, in 1992, the European Fighter Aircraft (EFA) Eurofighter Typhoon project was floundering, with Germany's threatened withdrawal causing Italy and Spain to have second thoughts, though Britain stated she would 'go it alone' if necessary.

Common sense eventually prevailed, and by 2006, in addition to the participating partners, Austria had purchased eighteen early F.1 air-defence variants (diverted from the RAF Tranche 1 order) and Saudi Arabia pledged to purchase forty-eight with an option for a further twenty-four. Unfortunately, an order from Greece for sixty was cancelled. In 2007, the overseas sales drive continued with a number of possible markets in the offing, including India. As a strong advocate of the Mirage III/2000 aircraft, in 2012 India confirmed its preference for the French Dassault Rafale to meet its future front-line aircraft requirements.

13

Mock-ups

As SBAC Farnborough International '94 approached (the thirty-ninth SBAC), expectations were high, but as it transpired, the show, which opened on 5 September, produced few great surprises. A number of new orders were announced, mergers confirmed, and projects discussed, but attendance by new types was disappointing. The Russians, who had dominated most of the Western air shows in recent years, brought only a comparatively modest contingent this year; in contrast, they were overshadowed by the strongest USA attendance for some time.

An alliance was formed between BAe, Eurocopter, and Rolls-Royce in connection with Britain's future combat helicopter, the first prototype (PT1) of the Eurocopter Tigre making demonstration flights. Meanwhile, a full-scale mock-up was placed on display by BAe. Here, too, the choice was split between the European Tiger (Tigre) with European-built engines—MTU, Turbomeca, and Rolls-Royce—a European-built anti-tank weapon system, the new long-range Trigat, or the purchase of the MDH AH-64 Apache, the Bell AH 1W Cobra, or the new Boeing RAH-66 Comanche. The latter was cancelled on 23 February 2004 after $6.3 billion had been spent on its twenty-year development programme. The Boeing AH-64D Longbow was the eventual winner; it was licence-assembled by Westland as the WAH-64D at Yeovil, in the West Country, in the UK.

As usual, the Russians dominated the combat aircraft displays, particularly when it came to Igor Votintcev in the desert-camouflaged Sukhoi Su-30MK Flanker and Viatcheslav Averianov in the canard-equipped Sukhoi Su-35. However, the Mikoyan MiG-29M Fulcrum and the re-designated MiG-33 export version did not appear as 'spirited' as their previous visits to the show.

British Aerospace took the opportunity to mark the Hawk's twentieth anniversary by flying fourteen black-painted Hawks, each displaying the identity of their overseas customers. This excellent idea was wasted as the type only made one flypast each day. Two of the BAe Hawk family were the newest members—the Hawk 200 (serial ZJ201) single-seat radar-equipped fighter in Royal Malaysian

The Eurocopter Tiger was a contender for the UK's British AAC attack helicopter requirement. (*Author*)

US Army MDH AH-64A Apache attack helicopters were deployed to Kosovo for training during the 1999 NATO-led campaign, but US President Bill Clinton prevented them from being used operationally. (*Author*)

The Longbow radar for the Boeing Apache (AH)-64D, sixty-seven of which were assembled by AgustaWestland for the British AAC-designated WAH-64D. (*AgustaWestland*)

The Sukhoi Su-30 Flanker gave rise to the Su-33 multirole aircraft ordered by the Indian and Chinese Air Forces, in addition to the platypus-nosed dual-seat Su-35. (*Author*)

The overseas sales of the BAe Hawk trainer were marked by fourteen RAF Hawk T.1As painted in all-over black, each with their overseas customer's name emblazoned. (*BAe sales brochure*)

The BAe single-seat, radar-equipped fighter—the Hawk 200. This is the development aircraft (serial ZJ201) with wing-tip AIM-9L Sidewinders. (*Air Power International*)

Air Force colours and the dual-purpose Hawk 102D (serial ZJ100) advanced combat LIFT trainer in the livery of the Royal Omani Air Force.

The fourteen Hawks depicted overseas sales to that date. The aircraft were pooled from Nos 208 (R) and 74 (R) RAF advanced flying training squadrons from No. 4 FTS RAF Valley, Angelsey, north Wales. The aircraft has been painted gloss black and the national flag and insignia of each of the Hawks' fourteen foreign air arms were applied to the wing, tail, and fuselage of each individual aircraft for a formation flypast known as the 'Python Formation'. Participating serials included:

Hawk T.1 XX250/1	representing 'United Kingdom'
Hawk T.1 XX261/2	'Finland'
Hawk T.1 XX282/3	'Kenya'
Hawk T.1 XX310/4	'Indonesia'
Hawk T.1 XX317/5	'Zimbabwe'
Hawk T.1 XX288/6	'Dubai'
Hawk T.1 XX314/7	'Abu Dhabi'
Hawk T.1 XX190/8	'Kuwait'
Hawk T.1 XX199/9	'United States' (MDC US Navy T-45 Goshawk)
Hawk T.1 XX218/10	'Saudi Arabia'

Hawk T.1 XX349/11	'Switzerland'
Hawk T.1 XX161/12	'South Korea'
Hawk T.1 XX202/202	spare aircraft
Hawk T.1 XX174/174	spare aircraft

Nine Hawk T.1s from the Red Arrows RAF aerobatic team were displayed, including serials XX227, XX233, XX237, XX253, XX260, XX264, XX266, XX294, XX306, and XX308.

Other main types displayed included:

Aero L-139 Albatros 2000 (f)
Aerospatiale ATR-72-200 (s) (first of type for British Airways in full BA livery)
Airbus Industrie A300-600F (s)
Airbus Industrie A320-114 (s) (first prototype)
Airbus Industrie A321-112 (f) (for Alitalia as I-BIXU)
Airbus Industrie A330 (f)
Airbus Industrie A340-311 (f) (first prototype, first four-engined Airbus)
Atlas/Denel Ace (f) (first prototype turboprop trainer)
The Atlas/Denel Ace

[(f) indicates flying, (s) indicates static display]

Later, the Atlas/Denel Ace turboprop crashed and was written off; the project was abandoned. The SAAF, which was due to purchase the aircraft, instead purchased the Swiss Pilatus PC.9 to replace its Atlas licence-built MB.336 Impala.

BAe Hawk Orders, Sales, and Exports

With a total of over 900 Hawks on order or delivered since the aircraft first entered service with the RAF in 1976, the Hawk has been established beyond doubt as the best aircraft of its class in the world.

Country	Model (Mk)	Quantity
United Kingdom	T.1	126 (including prototype serial XX154)
	T.1A	50
Finland	Mk 51/51A	50/7
Kenya	Mk 52	12
India	Mk 100 Series	60
Indonesia	Mk 53	20
Indonesia	Mk 109	8
Indonesia	Mk 209 (Batch 1)	16

Indonesia	Mk 209 (Batch II)	16
Zimbabwe	Mk 60	8
	Mk 60A	5
Dubai	Mk 61	9
Abu Dhabi	Mk 63A	16
Abu Dhabi	Mk 63C	4
Abu Dhabi	Mk 102	18
Oman	Mk 103	4
Oman	Mk 203	12
Kuwait	Mk 64	12
Saudi Arabia	Mk 65 (Batch 1)	30
Saudi Arabia	Mk 100 (Batch II)	20
Switzerland	Mk 66	20
South Africa	Mk 102	24
South Korea	Mk 67	20
Malaysia	Mk 108	10
Malaysia	Mk 208	18
Canada	Mk 115 (CT 155)	26 (Joint NATO Training Facility)
UK (RAF)	Mk 100 Series	28 (LIFT for Eurofighter—option for twenty-four)
USA	T. 45 Goshawk	187 (USN on-board carrier trainer)
Australia	Mk 127	36 LIFT
Qatar	Mk 100	12 *(cockpit compatible with Mirage 2000)

Status of UK-operated BAe Hawks *circa* 2007:

35+ written off
97 with RAF flying training units (eighteen of these used by No. 100 Squadron at RAF Leeming)
15 with Royal Navy Fleet Support Training Unit (FRADU)
10 with the Red Arrows Display Team, RAF Scampton, Lincolnshire
4 with BAe (trials/installation aircraft)
3 with QuinteQ at IAM/ETPS Boscombe Down, numbered '1', '2', and '3' on fins

In the new millennium, RAF BAe T.1/T.1A/1W Hawks are being refurbished by replacing the fuselage centre-sections to extend airframe hours, permitting their operational use until 2020. The rework (the fitment of Mk 65 centre-sections) takes twenty-one weeks per aircraft, with twenty-six reworks scheduled to be completed per year.

Farnborough '94 Participants

The participants at Farnborough '94 included:

BAe 748-2A-1 (s) CAA aircraft
BAe Hawk Mk 102D (f) 'Oman'
BAe Hawk Mk 200RDA (f) 'Malaysia'
BAe Sea Harrier F/A-2 (s) No. 899 NAS FAA (serial XZ455)
BAe Harrier GR 7 (s) No. 20 (R) Squadron RAF (air spare, serial ZD378/A)
BAe Harrier GR 7 (f) No. 20(R) Squadron RAF (serial ZD404/H)
BAe Nimrod MR.2P (s) No. 120 Squadron (serial XV254, displaying the Nimrod twenty-fifth anniversary badge)
BAe Jetstream 31 (s) (Jackie Stewart's Flying Scotsman, a private aircraft)
BAe Jetstream 41 (s)
BAe Jetstream 41 (s) (for Impulse Airlines, to become VH-AFR)
BAe Jetstream 61 (s) (BAe promotional aircraft)
BAC/Aerospatiale Concorde 102 G-BOAB in full British Airways colours (S) (public days only)
BAC 1-11-201AC (f) DRA
Bell 206B Jet Ranger III (s) (the 'round-the-world' flight record-holder)
Bell 206LT Long Ranger (f) (a police services rescue aircraft)
Bell AH-1W Super Cobra (s) (a US Marine Corps upgraded attack helicopter)
Bell OH-58D Kiowa (s) (a US Army 82nd Airborne Division armed scout helicopter)
Bell 230-CN (s) (a Chilean Navy demonstrator)
Bell 412CF(s) (Canadian Armed Forces)
Boeing CH-47D Chinook (s) (5/159th Aviation Co., US Army)
Canadair CL-601-3R Challenger (s)
Canadair SRS 100ER (s) (Air Canada, the first series production aircraft)
Cessna CitationJet (s)
Cessna Citation V Ultra (s)
Cessna Citation VII (s)
Cessna F-406 Caravan Vigilant (s) (French customs)
Cessna F-208 Grand Caravan (s)
Cessna Commander 114B (s) (advanced trainer)
Dassault-Breguet ATL 2 Atlantique (s) (French Navy)
Dassault Falcon 50 (s)
Dassault Falcon 900B (s)
Dassault Falcon 2000 (f)
Dassault Mirage 2000-D (s) (an Armée de l' Air multi-role fighter bomber)
Dassault Rafale C (s)
Dassault Rafale B (s) (both Rafale aircraft had their total number of flights annotated on their tail on a daily basis)

Dassault/Dornier Alpha Jet (s) (E-114/314-UC advanced trainer)
De Havilland Dash-8-100 (s) (City of Calgary)
Dornier 328 (f) (twin-engined regional jet)
Douglas DC3 (C-47A-40-DL) (f) (privately owned, registration G-BVOL, serial KG391)
English Electric P.1B Lightning (s) (privately owned, third prototype, serial XA847)
Eurocopter AS-555 Fennec (s) (ALAT French Army)
Eurocopter BK-117 (s) (Strathclyde Fire Brigade)
Eurocopter Tigre (f) (demonstrator attack helicopter)
FRA Cessna 441 Conquest (s)
Grumman Gulfstream IVSF (s)
Grumman Gulfstream IVSP (s) VR-48 US Navy
Grumman C-21A Goose (s) (privately owned)
Grob G115D Heron (s/f) (Royal Navy Flying Grading Flight)
GAF Nomad N24 (s) (Vickers Vimy replica support team)
Ilyushin Il-76LL Candid (s) (used as an engine test-bed, civil registration RA-76529)
Ilyushin Il-96MC (f) (registration 96000)
Ilyushin Il-103 (f)
Ilyushin Il-104 (f)
IAI Astra SP (s)
Lockheed C-5A Galaxy (s) (512th/436th AW USAF, serial 70-0460)
Lockheed P-3C Orion (s) (VP-24 US Navy)
Lockheed C-130H Hercules (s) (187th ALS Wyoming NG, serial 92-1535, City of Laramie)
Lockheed P-38J Lightning (f) (a Second World War bird from a private fighter collection in the US)
Lockheed F-16C Fighting Falcon (f/s) (23rd FS 52nd FW USAFE, serial 91-0416)
McDonnell-Douglas AH-64A Apache (s) (2nd Battalion 1st Aviation Regiment US Army)
McDonnell-Douglas AH-64D Longbow Apache (s) (MDH demonstrator)
McDonnell-Douglas F-15E Strike Eagle (s) (492 FS 48th FW USAFE, serial 90-0248)
McDonnell-Douglas F/A-18C Hornet (f) (US Navy VFA-82 USS America CVW-1)
McDonnell-Douglas F/A-18C Hornet (s) (US Navy VFA-86 USS America CVW-1)
McDonnell-Douglas C-17A Globemaster III (f) (437th AL Wing USAF, serial 92-3294)
McDonnell-Douglas MD900 Explorer (s) (the first prototype)
Mikoyan MiG-29S Fulcrum (f) (MiG-MAPO demonstration aircraft)
Mikoyan MiG-29M (s) (Fulcrum MIG-MAPO demonstrator)
Mil Helicopters Mi-17M Hip (s) (Kazan Helicopters)

Myasischev M-55 Mystic B (f) (demonstrator)
Northrop F-5B Freedom Fighter (s) (338Skv Royal Norwegian AF F-5 MLU demonstrator)
Panavia Tornado F.2 TIARA (s) (DRA serial ZD902)
Panavia Tornado GR.1A (s) (No. 2 Squadron RAF, serial ZA404, 'GR 4 titles')
PBN BN-2T-4S Islander MSSA (s) (Westinghouse APG-66 nose radar, serial G-BVHX)
PBN BN-2B-26 Islander (s)
PBN BN-2T-4S Defender 4000 (s) (company demonstrator)
Pilatus PC-7 Mk II (s) (company demonstrator)
Pilatus PC-9 (f) (turboprop advanced trainer, company demonstrator)
Pilatus PC-12 (XII) (f/s) (company demonstrator)
Piper PA-31-350 Chieftain (s) (Hunting Aviation)
Raytheon Hawker 800 (s) (BAe/Raytheon 125-800 demonstrator)
Raytheon Hawker 1000 (s)
Raytheon 125-1000 (demonstrator)
Robinson R-22 (s) (Sloane Helicopters)
Robinson R-44 Astro (s) (Sloane Helicopters)
Saab 340 AEW Tp.100 (f) (for Swedish AF, serial 100002)
Shorts 360-300 (s) ('Atlantic Reconnaissance')
Sikorsky UH-60A/C Black Hawk (s) (US Army serial, 86-26040)
Sikorsky SH-60FSeahawk (s) (US Navy on loan to Sikorsky, serial 164802)
Sikorsky HH-60 Pave Hawk (s) (USAF serial 86-26112)
Sikorsky S-76A Spirit (s) (company demonstrator—medevac fit)
Slingsby T-67M-260 Firefly (f) (Slingsby demonstrator)
Socata/Aerospatiale TBM-700 (f/s) (company demonstrator)
Socata/Aerospatiale TB-31 Omega (f) (company demonstrator
Socata/Aerospatiale TB-30 Epsilon (s) (company demonstrator)
Socata/Aerospatiale TB-20 Trinidad (s) (company demonstrator)
Sukhoi Su-27 Flanker (f)
Sukhoi Su-29LL (f)
Sukhoi Su-30MK Flanker (f)
Sukhoi Su-31 (f) (Breitling aerobatic, private)
Sukhoi Su-35 Flanker (f) (the third prototype)
Tupolev Tu-204 (f/s) (Vnukovo Airlines, carried both Russian and English titles)
Valmet L-90TP Redigo (f/s) (Valmet demonstrator)
Vickers FB-27 Vimy (f/s) (Shell-sponsored replica, serial NX71MY)
Westland Lynx HAS 3 (s) (No. 702 NAS, serial XZ730)
Westland Wessex HC.2 (f) (No. 22 (SAR) Squadron RAF, serial XT604)
Westland/EHI EH-101 (s) (development aircraft PP.9)
WSK-PZL-Mielec M-93V (f) (demonstrator)

The following aircraft flew on trade days:

Dassault Falcon 20DC G-FRAP FR Aviation (also public days)
Dassault Falcon 20E G-FRAD FR Aviation (also public days)
Dassault Falcon 20DC G-FRAO FR Aviation (also public days)
Dassault Falcon 20DC G-FRAK FR Aviation (also public days)
Panavia Tornado GR.1 ZA326 DRA (Boscombe Down—Raspberry Ripple)
BAe Nimrod MR.2P XV229 (RAF Kinloss Wing—all-over hemp camouflage)
BAC 1-11 XX105 DRA

Following its British public debut at RAF Mildenhall, Suffolk, the USAF MDC C-17A Globemaster III made its second appearance at Farnborough and put on an impressive display routine, with a fairly noisy high-speed 'reverse' down the runway before getting airborne after a surprisingly short take-off roll. At this time, no one knew that the RAF would also operate the type.

Admirers of veteran 'flying warbirds' were well-served at the show. Of particular interest was the immaculate Lockheed P-38J Lightning (which the RAF had rejected during the Second World War due to dispute with Lockheed) flying past in the company of an LM F-16C Fighting Falcon. However, probably the most impressive display was that of the Russian Sukhoi Su-35 advanced multi-role Flanker. The aircraft suffered a pre-show mishap when it overshot the runway and damaged a landing light, but following a hasty repair it stole the show with its stunning 'cobra' tail-slide manoeuvres.

The five-crew Lockheed C-5 Galaxy strategic transport was nicknamed the 'Georgian Giant' by USAF personnel. (*Warplanes collection*)

The following aircraft flew only at Farnborough '94's public day heritage display:

Manufacturer	Type	Serial	Operator	Remarks
Avro	Lancaster B.1	PA474	BoBMF	No. 9 Squadron markings
Bristol	Bolingbroke IVT	Z5722	IWM Duxford	No. 68 Squadron (G-BPIV)
DH	Tiger Moth	T6818	Shuttleworth	G-ANKT
DH	Chipmunk T.10	WD331	6 AEF RAF	
DH	Chipmunk T.10	WB550	6 AEF RAF	
DH	Vampire 215		private owner	
Douglas	DC-3 Dakota	G-AMPO	Air Atlantique	
Douglas	DC-6 A/B	G-SIXC	Air Atlantique	
EE	Canberra TT.18	WJ680	Canberra Flight	No. 100 Squadron (G-BURM)
Fairey	Swordfish	W5856	RNHF	No. 810 Squadron markings
Fairey	Swordfish	LS326	RNHF	
Folland	Gnat T.1	XR991	Intrepid Aviation	G-MOUR
Gloster	Meteor NF.11	WM167	Jet Heritage	No. 141 Squadron G-LOSM
Hawker	Fury FB.10	N36SF	private owner	RAN colour scheme
Hawker	Hunter T.8	XL573	private owner	
Hawker	Hurricane IIC	PZ865	BoBMF	No. 303 (Polish) Squadron
Hunting	Jet Provost T.1	G-AOBU	private owner	
Miles	M.14 Hawk	P6382	Shuttleworth	16 EFTS (G-AJRS)
Miles	Messenger	RG333	private owner	
NA	Harvard IIB	FS728	private owner	G-BAFM
Percival	Gull 5	G-ADPR	Shuttleworth	
Percival	Prentice T.1	VS610	private owner	G-AOKL
Percival	Provost T.1	XF597	private owner	G-BKFW
SA	Bulldog T.1	XX553	London UAS	
Supermarine	Spitfire PR.XIX	PS915	BoBMF	SEAC markings
Supermarine	Spitfire T.IX	ML407	Carolyn Grace	No. 485 Squadron markings

In addition, various 'mock-ups' were situated around the display site, including a Eurocopter Tigre with British Army titles and roundels, a BAe Hawk 200 single-seat fighter, an Airbus Future Large Aircraft FLA (Airbus A400M), and a Eurofighter 2000 (EFA) Typhoon. The Typhoon was displayed inside a demonstration dome that saw attendees queuing for hours; spaces were limited, especially for the excellent air combat sortie video, which was projected onto a large screen with surround-sound. This was a very impressive BAe promotion for an aircraft that had not even taken to the air at this time.

Farnborough '94 was shorter than previous shows, lasting for seven days instead of eight. This was welcomed by the exhibitors, but the enthusiasts, who traditionally arrived on Friday, missed out. Nonetheless, it was still a good day out, with ballooning, aerobatics and vintage airliners in the morning and a major flying display in the afternoon. Without doubt, the major event of the weekend was the much-publicised departure for Australia of the Shell-sponsored replica Vickers Vimy, alongside its twin-engined light transport GAF Nomad support plane.

Meanwhile, Airbus muted its proposal for its A3XX double-decker Super Jumbo, which was intended to carry up to 850 passengers. At the time, no decision had been made as to whether to go ahead with the project. Airbus was waiting until some of the major carriers 'firmed up' their initial enquiries for the giant airliner.

In the static park, three contenders were exhibited for the RAF's maritime-patrol aircraft replacement—the existing BAe Nimrod (to be upgraded to the '2000' standard), an upgraded Lockheed P-3C Orion, and the Dassault Atlantique 2. The Atlantique 2, offered with Allison engines, was considered to be the outsider. The Nimrod eventually won the day with eighteen (later reduced to twelve) to be upgraded to MRA 4 standard, with six due to be in service by 2010. However, this order was cancelled in 2012 by the UK coalition government's Defence Review.

Even though Saab had five JAS 39 Gripens that had accrued some 1,400 hours of flying between them, the tight development programme precluded the type's appearance at the show, although the prototype Saab 340 AEW aircraft was displayed, equipped with the Ericsson Erieye radar.

Boeing announce their latest B.737 development, the 164–189-seat 737-800, which was intended to be a competitor to the Airbus A321. The first aircraft was scheduled to be delivered in March 1998, with forty firm orders placed by four customers.

Airbus displayed its A330 for the first time at the show in the new colours of Cathay Pacific. For the first time, Airbus Industrie had taken the lead, with sixty-nine commercial aircraft ordered in the first half of the year compared to Boeing's fifty-four. Customers could choose between the A320, A321, A319, and A330 twins or the four-engined A340-200 or -300. Boeing offered a choice between the four-engined 747 and the new -777 twin—two notably different machines, but both wide-bodies.

With 1994 marking the twenty-fifth anniversary of Concorde, Tupolev interestingly announced that it was taking one of its Tu-144LL Charger SSTs out of storage to prepare for its use under contract to the USA as a flying test-bed for research work on a new generation of SSTs. A research agreement for a joint project had already been signed in April by Aerospatiale, British Aerospace, and DASA, and engine manufacturers (particularly Snecma and Rolls-Royce) were working on the power plant design at this time. Concorde made its appearance at Farnborough on the public days, 10 and 11 September.

Years of prevarication, technical delays, and difficulties—not least with the Allison engines—meant that the deliveries of the Airbus FLA (Airbus Military A400M) to the *Armée de l'Air* did not commence until 2012. (*Airbus Military brochure*)

The Dassault Atlantique 2 (the upgrade of the Dassault Breguet Atlantic 1) failed to win any foreign orders for the French company. (*Dassault*)

Boeing's 777 twin wide-body with full ETOPS. The airliner is resplendent in the company's new corporate colours. (*Author*)

Embraer continued to promote its thirty-seat EMB-120 Brasilia twin-turboprop short-haul commuter. The aircraft incorporated a number of changes, including improved noise reduction, a redesigned interior, increased cargo capacity to 700 kg (1,543 lb), and an improved flap system. The company developed two new military variants of this airframe for the Amazon Surveillance System. Five EMB-120EW early warning aircraft were fitted with Ericsson Erieye radar and three EMB-120SRs were equipped with Synthetic Aperture Radar (SAR), TV/FLIR, and a communications system for monitoring illegal communications traffic. The company's advanced ALX attack version of its Tucano trainer, the EMB-312H Super Tucano for the Forca Aerea Brasileira, had structural and systems modifications. Available in single-seat and two-seat configurations, the wing had five hardpoints for stores and new equipment including multi- function cockpit displays, HUD, GPS, and provision for FLIR, RWR and a chaff dispenser.

On the regional transport side, the ATR consortium Aerospatiale/Alenia continued its success with orders from Air Tahiti, Vietnam Airlines, and TACV-Cabo Verde Airlines. Their competitors also saw more orders, with interest in the Saab 2000, the Fokker 70, and the Canadair Regional Jet.

Like the Airbus, the Avions Transport Regional (ATR) models were available in an extended family, with still more to come. The first of them, the ATR 42-500, was undergoing flight tests at Toulouse. At this time, the carriers' opinions were still divided between the merits of an economical twin turboprop and those of the speedier but more costly twinjet. In the case of the BAe, the decision was between the twin turboprop and the company's four-jet 146, which had been repackaged as the RJ Series under the Avro banner at Woodford. By the end of the decade, the 146 had disappeared from the scene completely; however, it should be said that in 2006 some of the total build remained in service with many smaller operators around the world—some converted to freighters. One hundred and fifty-eight of the 209 BAe 146s and 152 of the 166 RJs delivered remained in use, many under lease agreements with BAe Systems, the aircraft having previously been brought out of desert storage at Davis Monthan AFB in the USA.

Polish PZL's I-11 Iryda (registration SP-PWE) was the fifth prototype of the company's new jet trainer designated M-93V. It was powered by two Rolls -Royce Viper 535s in the hope of attracting Western customers.

The Czech aircraft manufacturer LET revealed it had built 1,084 Walter M-601E-powered L-410 Turbolet twin-turboprop short-haul commuter and utility aircraft since the type had first entered service with Slov-Air in late 1971. It had also built six L-610M and one L-610G airliners, with the L-420, an improved version of the L-410, under development.

A glimpse into the future was provided by the first flight of the German prototype Eurofighter 2000 on 29 March 1994. This was followed by the second aircraft, this time in the UK, on 6 April. In a special Eurofighter 2000 display at Farnborough (adjacent to its presentation park), British Aerospace incorporated a very impressive technology exhibition, with a full-size single-seat Eurofighter 'mock-up' and the Eurofighter 2000 mission demonstrator. The demonstrator reproduced new air-to-air 'dog fight' sequences with a revealing 'twist in the tale', assisted by a 180-degree wraparound screen with surround sound.

BAe's comprehensive range of commercial aircraft was on display in the static park, marking the first appearance of the new seventy-seat Jetstream 61, the twenty-nine-seat Jetstream 41 (for Impulse of Australia), and a Jetstream 31 corporate air shuttle completing the turboprop line up. Meanwhile, Avro displayed

Following the successful first flight of the EADS Airbus A380, the company ran into difficulties with the type due to cable loom problems. The cost of delivery delays and rework means that it is estimated it will take fifteen years before the company begins to see a profit on its 'Super Jumbo' investment. (*Air Power International*)

its RJ70 Avroliner that was specially commissioned for Air Malta, with their line up completed by an Airbus A320.

Unfortunately, in 1994 BAe's civil operations remained depressed—a situation from which the company never fully recovered. Despite the success of previous years, when Sir Frederick Handley-Page's Jetstream and Sir Geoffrey de Havilland's DH 125 Dragon business jet had appeared to get things right, BAe's demise on the civil scene could not be averted. The company had already sold its subsidiary corporate jets operation at Hatfield and the production of its popular -125/-1000 series of business jets to Raytheon at Kansas, in the USA.

The year 1992 was a good one for the BAe 125 business jet, with orders for thirty-two examples materialising out of the fifty that were predicted earlier in the year. That order represented nearly 50 per cent of the market for medium-sized cabin aircraft. Sales were particularly strong in Europe and Africa; these non-North American sales were the highest for nearly thirty years. A total of thirty 125-800s and -1000s were delivered in 1992, with a similar amount delivered in 1993.

Included amongst the orders was the first sale of a BAe 125 aircraft to the Confederation of Independent States (CIS)—formerly the USSR—and the acquisition of a BAe 1000 by the government of Turkmenistan. The CIS sale was soon followed by the formal granting of Russian certification to the 1000 series—a vital step to selling aircraft into the new Russian republic. The approval meant that the -125 and -1000 series could be registered in CIS countries and would qualify for cheaper navigation and airfield charges. Without such certification, business jets have to pay navigation charges at Western levels.

Another significant order at this time was made by Executive Jet Aviation of Columbus, Ohio, who requested fourteen Hawker 1000s in a deal worth over $250 million. The original Executive Jet programme involved 111 owners with twenty-five Cessna Citations.

June 1993 had seen the worldwide -125 fleet flying hours exceed 4 million; the type had accumulated a valuable reservoir of operational experience. In the same month, BAe announced that it intended to sell the business jet facility to the American Raytheon Company, maker of the Beech range of aircraft in the USA. The agreement stated that the manufacture of the 125 at the Chester plant would continue for three years, when the situation would be reviewed. The 125 was renamed as the 'Hawker', and all subsequent production of the aircraft was undertaken in the USA.

Raytheon approved and funded upgrade programmes for both the Hawker 800 and the Hawker 1000. The work included the preparation of the first of twenty-seven Hawker 800 Special Mission aircraft for the Japanese Air Self-Defence Force (JASDF). Designated U-125A, the aircraft were intended to be used for search-and-rescue duties and were fitted with 360-degree-scan radar and a forward-looking infra-red system. They were also capable of dropping marker flares and rescue equipment. The twenty-seven aircraft were delivered between 1995 and 2003.

The BAe Jetstream family continued to sell, but at a much reduced pace compared to former years. Sales of the smallest member were the most successful—the Jetstream 31 and Jetstream Super 31 approached the 400 mark worldwide, and the fleet flying hours amounted to 4 million, with 158 Jetstream 31/32s still in use in 2006 (most in North and South America). Deliveries of the stretched and much-enhanced Jetstream 41 were well underway, with orders reaching thirty-nine for five customers in the UK, USA, Australia, and Europe. The final total of aircraft delivered was 106. Two BAe Jetstream developments were announced in 1993, the 51 and 71, but they failed to materialise.

Overall sales of the ATP, a BAe development of the successful HS 748, were disappointing, with an eventual total of sixty-four delivered. The main customer for the type was British Airways, which at one time operated thirteen on regional routes—particularly in Scotland and the Highlands.

On the regional jet scene, the former family of BAe 146 regional airliners were now marketed under the Avroliner banner. The new airliners were called 'RJs' (Regional Jets) and embodied up-rated engines and all-digital avionics, plus other minor modifications and improvements (especially to the interior). The smallest of the range was the Avro RJ70, which seated seventy to ninety-four passengers. This aircraft had a shortened fuselage and de-rated engines, and it was first developed as direct competition to a regional aircraft with similar capacity, such as the Fokker 70.

The Avro RJ85 seats eighty-five to 112 passengers and was the first of the new series to be delivered to Swiss airline Crossair in April 1993. The largest model, the RJ100, seats 100 to 128 passengers and is the same size as the RJ115, with mid-cabin exits, bigger air-conditioning units, and higher design weights. The RJ115 seats 116 passengers in a six-abreast configuration. The prefix 'RJX' was assigned to a newly proposed twin-engined development that was scheduled to enter service at the turn of the century; it never materialised. Instead, BAe Systems ceased civil airliner production completely.

By early 1994, the Pilatus Britten-Norman family of PBN Islanders had notched up 10 million flying hours—a feat achieved by less than a handful of British airliners. It had taken twenty-seven years for the Islander to reach this milestone, and during this time it established a record for the number of British aircraft of one type sold. The total sales amounted to 1,270 aircraft, of which more than 800 remained in use in 2006. The average use of the aircraft amounts to 550 hours per year. The lead aircraft, serial No. 602, in service with Sunflower Airlines in Fiji, clocked up well over 28,000 flying hours by 1994. This is equivalent to being airborne continuously for over three years. The latest variant, the BN26-4S Defender 4000 first took to the air (registered as G-SURV) on 17 August 1994; however, by 2013, it had still failed to secure any orders.

14

The Airbus A3XX 'Super Jumbo' is Revealed

Farnborough International '96 was generally deemed to have been more successful than the 1994 show, with many companies announcing orders and the SBAC, the organisers of the event, reporting a greater take-up of chalets. Additionally, an increasing number of firms were booking new stand space. Gone was the familiar blue-and-white-striped roofing of the terraces of yesteryear; they were replaced by De Boer Structures' new modularised hospitality suites. By now, of course, Farnborough had to compete on an international basis with Paris (on a biennial, rotating basis), plus Moscow, Berlin, the emerging Asian events (such as those in Indonesia and Singapore), Dubai (in the Middle East), and those in Australasia.

To the companies involved in the aerospace industry, these myriad events left plenty of choice, if not a constant drain of their often-limited advertising and sales marketing budgets. As such, many were unwilling to send all of their products to every show. This was noticeable at Farnborough '96 by the significant reduction in the number of airframes actually on show; it was the lowest number since the 1960s, when the show had become 'international'.

Nevertheless, more than 100 of the world's latest military and commercial types were displayed. Amongst the latter were the duet from Airbus, the A330 and A340; these aircraft produced a tight display that had no equal among such large aircraft, with Boeing's rival 777 prototype remaining firmly on the ground.

Improving prospects in the civil sector led to Airbus revealing that it was resisting pressure from Boeing (which was close to launching its own stretched 747-500/600 Jumbo Jet), keeping to its original plan of launching its 'Super Jumbo' A3XX (A380) 'some time in 1997'. The consortium stated that the project 'would not be rushed', and scheduled the aircraft to enter service in 2003. In 2007, the company estimated that production problems with type's cabling would delay its introduction until 2009.

The A380 is the world's largest airliner, and it is the first with twin-aisle seating in both the upper and lower decks. Airbus was convinced that the VLA (Very Large

Aircraft) would be the only way of carrying more passengers without necessitating the building of new airports or extending those that were already suffering from congestion. It planned that the growth versions of the 'Super Jumbo' would take off with more than 1,000 passengers on board.

In contrast, Boeing shelved their large aircraft project in favour of their fast point-to-point Sonic Cruiser, which was subsequently supplanted by the 787 Dreamliner. The Boeing 747 stretch was later resurrected again to challenge Airbus, who reciprocated by introducing their A350 Xtra-wide-bodied challenger to the Dreamliner. As the quest for overall dominance of the civil passenger market goes on, will it be Europe or the USA that is the eventual winner?

Unfortunately, the problems facing Airbus and its parent company, EADS, have continued. In addition to the A380's existing problems, it was grounded due to 'wing cracks'; meanwhile, there was a forced redesign of the A350, which was originally based on the A330. In this sense, it was likely to be Boeing who had got it right again. The company's Dreamliner was built almost entirely of composite materials; the fuselage section represented the largest composites ever built for a commercial airliner. However, on 9 January 2013, three ANA Japan Airlines Boeing 787 Dreamliners were grounded—one due to fuel leaking into the cabin, one with a cracked cockpit window, and one due to smoke from an under-cabin fire that was traced to faulty lithium batteries (used to provide the airliner with electrical power on the ground). The latter incident resulted in the grounding of all fifty 787s delivered up to late January 2013, as Boeing engineers worked to find solutions to the problems.

Back in the 1990s, Airbus stated that the A380 would fit within an 80 metre × 80 metre box to enable it to be accommodated by the new 80-metre-wide departure gates planned for new airports and terminals. The wing would be around 40 per cent larger than that of the Boeing 747-400, and primary structures in carbon-fibre would include the wing out-board of the outer engines, the fin, and horizontal tail-plane. Considering that the tail-plane alone would be of a similar size to the entire wing of the Airbus A310, the big challenge for the A380 would be determining the dimensions of these structures.

The A380 cockpit was designed to have full commonality with the existing A320/A330/A340 range so that operators could benefit from full cross-crew qualification across all modern Airbus types. The aircraft would also have a variable-camber wing built by BAe Systems at Broughton, in the UK. In 2006, BAe Systems sold their 20 per cent share in Airbus to their parent company, EADS.

Development of the A380 initially represented an $8 billion investment for Airbus, though by October 2006, with manufacturing problems associated with the wiring looms and delivery dates to launch customers revised, claims for compensation were likely to amount to £210 million. Emirates, for example, were claiming £150 million in compensation after delivery of their first aircraft was postponed for the third time. The Dubai-based company had ordered forty-five of the 'Super Jumbos', making it the leading launch customer. Other airlines, including Virgin Atlantic and Qantas, were also expected to claim for late delivery.

Military flying dominated the show. It was Yevgeny Frolov, in the big thrust-vectoring Sukhoi Su-27 Flanker, who took the honours for the most dramatic display sequence, with the moveable nozzles enabling what was described as the 'Super Cobra' to be flown. While the big Russian Sukhoi thundered and tail-slid around the sky, its western counterparts struggled to fight it out for second place. That position ultimately went to the Eurofighter, which finally made its debut in at the show; its display was rather sedated due to underdeveloped flight-control software. Even so, Deputy Prime Minister Michael Heseltine reaffirmed the United Kingdom's full commitment to the project in his opening address.

In 1996, the UK Police Aviation Services ordered ten MDH MD900 Explorer helicopters—the largest single purchase of the type at that time in Europe. The first eight-seat helicopter was due for delivery in June 1997 to enter service with the Wiltshire Constabulary Air Support Unit. The helicopters were to be used for police work and air-ambulance services. Three more were scheduled for delivery in 1998, with the balance to follow over subsequent years. The UK order took Explorer sales to around seventy. In mid-July 1996, the European Joint Aviation Authority's certification of the Explorer had been ratified by eleven European agencies, including the UK's CAA. Additional international certification included Japan and South Korea. The type was certificated by the Federal Aviation Authority (FAA) in the USA in December 1994.

London's policemen and those in twelve other English counties are also equipped with helicopters, with more than 400 operational (around 30 per cent for off-shore operations). The Metropolitan Police Air Support Unit is based at Lippitts Hill at Loughton, north-east of London, in huts dating back to the early 1940s, when the place was a prisoner of war camp. It has been operating a mini taskforce of twin-engined rotary-winged types since 1980.

All the machines are equipped with sophisticated communications systems, plus forward-looking infra-red FLIR/TV cameras, high-intensity 30-million-candela searchlights, loudspeakers, and so on. Over a fifteen-month period, three of the Air Support Unit's ASU helicopters enabled 469 arrests to be made—an average of 1.5 per day. They also helped to find over 100 missing people and saved the lives of three individuals.

Operating day and night with three-man crews—a pilot, a police navigation officer, and an observer—the ASU works on a shift system. The helicopter operates out of Lippitts Hill in the mornings and out of a secondary base at Fairoaks, situated south-west of London, on the afternoon and night shifts. Two machines remain on alert on each shift, never more than seven or eight minutes away from central London or Surrey (whose force is also able to call on the Met ASU). On average, the unit logs between eight and ten one-hour flights per day, with two or three missions accomplished during each sortie. The helicopters, which maintain constant radio contact with Heathrow, can refuel at any one of four points around Greater London.

A little-known fact is that every one of London's police officers, all of whom carry radios, can call on the ASU for help. The Met's powerful mainframe computer records all such calls and relays them to Lippitts Hill, where the ASU dispatcher checks the messages on a terminal and decides whether or not to launch a mission.

This is a summary of UK police turbine helicopter types operated since the 1980s:

Agusta 109 Power
Bell 206/Agusta-Bell 206 Jet Rangers
Bell 206L Long Ranger
Bell 212 (offshore)
Bell 214 (offshore)
Bell 222
EH 101 Merlin
Eurocopter AS 332 Super Puma (offshore)
Eurocopter AS 350 Ecureuil
Eurocopter AS355 Ecureuil
Eurocopter AS 365 Dauphin (sixteen in total, fifteen offshore)
Eurocopter BK 117
Eurocopter BO 105 (twenty-two in total, ten offshore)
Eurocopter SA Lama 315
Eurocopter SA Gazelle 341
MDC/Hughes 500
MDC/Hughes 520N
Mil Mi-2
Sikorsky S.61
Sikorsky S.76

The award for the most disappointing and boring participant at Farnborough must go to the American Boeing B-2A Spirit stealth bomber. The aircraft made only a fleeting visit on Monday's press day, spending less than five minutes in front of the crowd. On top of this, the aircraft was late due to the weather; the residents of nearby Aldershot probably saw more of the aeroplane, as it circled to the south of the show area, than those eagerly awaiting its arrival in the main arena.

The 393rd Bomb Squadron B-2A (serial 88-0332, tail code WM *Spirit of Washington*) had left Whiteman AFB, Missouri, as 'Spirit 71' at midnight local time on 1 September for a medium-level (25,000-feet) flight across Maine and Canada, the North Atlantic, Shannon, and on to the show. The crew pilot, Capt. Daryl Davies, and mission commander, Capt. Frank Cavuoti, performed just two flypasts in front of the crowd before they headed back to Missouri over the same route. The two flypasts were identical—straight and high. A planned simulated bomb run at the Vliehors Range during the return flight was cancelled. It made one wonder, if the twenty-one-hour return flight was really worth it. However, in hindsight, it probably was good practice for the nineteen-hour missions flown by the type during the Kosovo War in early 1999.

During its flight from the USA to Europe the billion-dollar stealth bomber made four air-to-air refuelling contacts from Boeing KC-135R tankers and covered around 15,000 track miles. The Farnborough display aircraft landed back at Whiteman at 9 p.m. local time on Monday 2 September, after completing its twenty-one-hour 'Global Power' mission.

A show-stopper in 1994, the Vickers Vimy replica (civil registration G-EAOU) returned to Farnborough in advance of the recreation of another epic long-distance flight, taking its place alongside the fastest airliner in the world—British Airways Concorde registration G-BOAC—just to prove that size doesn't matter.

The Vimy was due to set off for South Africa in 1997 to recreate the 1920 competition to be the first aircraft to reach the country from the United Kingdom, following the route taken by Lt-Col. Pierre van Ryneveld and Maj. Christopher Quintin-Brand of the South African Air Force during their epic six-week flight that had begun at Brooklands, Surrey, on 4 February. In the course of their journey, they wrecked two Vimys; they eventually arrived in Cape Town in an Airco DH.9 on 20 March 1920, after 109 hours and twelve minutes of flying. In 1997, the replica Vickers Vimy was owned by Peter McMillan, who was accompanied on the epic recreation flight by Ken Snell, who was a British Airways Concorde pilot.

The major Russian participants at Farnborough '94 produced a number of news stories as well as being the star performers during the flying displays. Amongst these was the thrust-vectoring Sukhoi Su-37, whose daredevil display took some clearing before being given the go-ahead by Farnborough's flying control committee. The McDonnell-Douglas F/A-18 Hornet also encountered problems with the committee, having allegedly infringed mandatory minimum-height limits during one of its displays. While the 'big' Sukhoi performance was outstanding, it should not be forgotten that it performed 'clean'; it is still open to debate how it might have performed in a 'dog fight' against the diminutive Lockheed Martin F-16 Viper. There is little doubt, however, that the Sukhoi Su-27, Su-30, and Su-37 Flanker family are formidable aeroplanes, and therefore it is of little surprise that the Russian government are making the most of its export potential. However, the agreement to permit China to licence-build the type has caused much concern, not least in Russia, as it is not beyond the realms of possibility that the Chinese will produce their own derivative—as they have done many times before—to secure some of the lucrative Asian military combat aircraft market for themselves.

Among the other Russian performers in the flying display was the Mikoyan MiG-AT, the new tandem two-seat trainer aircraft—a potential contender for the Russian Federation Air Force's future trainer needs. This aircraft was first shown at the Paris Air Show in 1995, but it had not flown since that time until early in 1996.

Its main rival, the Yak/Aermacchi Yak-130, was not present at Farnborough. Both types were eventually selected for service with the Russian Federation Air Force (RFAF) in the new millennium.

The Soviet Kamov Ka-50 Hokum army attack helicopter that appeared at the Farnborough '94 static display gave its first flight demonstration in 1996.

There was a multinational array of combat aircraft on the flight line at Farnborough '96, including: a Lockheed Martin F-16C Fighting Falcon; Panavia Tornado GR.1s; a dual-seat MDC F/A-18D Hornet; a Sukhoi Su-37 Flanker and Su-30 Flanker; the Spanish two-seat Eurofighter 2000, which made its first flight on 31 August; and the BAe Hawk 100.

On the civil scene, the possibility of a Russian buy-out of the failed Dutch aerospace firm Fokker was still on the table. The company had been kept going just to complete outstanding orders for its civil types. In addition to Yakovlev's interest, there had been speculation about a possible buy-out by the South Korean industrial giant Samsung. However, in the end, Fokker virtually disappeared from the scene—although not completely, and in 2006 there was talk of reviving its regional jet designs.

Tupolev promoted its Tu-204 series at the show. The type had been developed into a family of airliners with various interior and engine options, thirty of the production aircraft to be powered by Rolls-Royce engines. Tupolev also announced plans for several new types and derivatives, including the Tu-334 short and

Above: This General Dynamics F-16C was presented in its upgraded form as the Lockheed Martin Viper, its original name before 'Fighting Falcon' was settled on. (*Author*)

Below: The Yakovlev Yak-130 has been developed in conjunction with Aermacchi of Italy, but only the RFAF has ordered the type. (*Author*)

Opposite: The Mikoyan MiG AT is the first indigenous jet trainer to be produced by the Russian aircraft industry. (*Author*)

The original MDC F/A-18A Hornet re-emerged as the Boeing F/A-18C. (*Author*)

An RNZAF BAC Strikemaster. (*Air Power International*)

medium-range airliner designed to replace the long-standing Tu-134, the Tu-330 cargo plane, and the Tu-324 regional transport. Tupolev also stated that its Tu-34 light aircraft would eventually enter production.

An indication of greater cooperation between Russian and Western aerospace companies was given by the announcement at Farnborough '96 of a major deal between the EOS Logistics Group (of Britain, France, and Germany) and the Russian company Ilyushin to create fully computerised manuals and logistics-support materials to Western standards for Ilyushin aircraft.

A significant talking point at the show was the possibility of a replacement combat 'western' type for the new Czech Air Force and Air Defence Command, and Aero was promoting its L-59 and L-159 combat derivatives of its highly successful L-39 Albatros series. A number of companies were already vying for the country's Mikoyan MiG-21 replacement, including: Lockheed Martin, with its F-16 Fighting Falcon; MDC's F/A-18 Hornet; and Saab, with its JAS 39 Gripen. It was alleged that Hungary was also interested in the Gripen. Both countries subsequently chose this excellent Swedish fighter to fulfil the air combat needs of their air forces for the new millennium.

The principal Romanian aircraft and helicopter manufacturer, IAR-Brasov, was keen to promote its links with Bell Helicopters, stressing that the Romanian Armed Forces intended to acquire the Bell AH-1 Cobra attack helicopter and that the company would undertake licence-production of the type. The company also promoted its licence-produced IAR-330 Puma tactical transport helicopter. Another significant Romanian company, Aerostar, was offering its Lancer MiG-21 upgrades in conjunction with Elbit of Israel. The upgrade produced an excellent aeroplane that was comparable to an early General Dynamics F-16A Fighting Falcon with a completely new cockpit and avionics, in addition to various other upgrade features that effectively 'Westernised' the aeroplane. The contract to upgrade the entire fleet of Romanian Air Force aircraft was signed in 1993, with the first upgraded aircraft flying in 1995. One of the initially converted and updated Lancers, a single-seat version, was displayed at Farnborough '96.

Romaero Airstar were promoting the Airstar 2500, a derivative of the BAC 1-11. However, following the transfer of the whole programme to Romania, only a handful of Rombac 1-11s were completed. Nevertheless, Airstar endeavoured to foster interest in its proposed upgraded development of the outdated British airliner, but without much success.

There were a large number of orders for new airliners from the established manufacturers. For example, the Brazilian company Embraer concluded a contract with Continental Express for twenty-five firm orders for its EMB-145 regional airliner, with the airline taking options for a further 175. The sale brought firm orders for the EMB-145 twin-jet to sixty-three, with 217 options.

Boeing announced the go-ahead of the 757-300 with twelve firm orders and twelve options from the German carrier Condor; the series 300 was to be

The Czech Aero L-39 Albatros served for the latter part of the Cold War as the main jet trainer for the Soviet VVS and their Warpac partners, replacing the defunct Mikoyan MiG-15UTI. (*Author*)

The Aero L-159 Albatros is a combat-capable development of the L-59 Albatros. (*Author*)

The IAR Lancer is a Romanian upgrade of the ubiquitous Soviet Mikoyan MiF-21 Fishbed. (*IAR brochure*)

7.1 metres longer than the 200, although the Condor aircraft had a 252-seat layout. The order was part of $8.1 billion worth of contracts, with the Australian leasing company Ansett Worldwide ordering ten Boeing 737-700s, six 737-300s, and nine 767-300ERs. The International Lease Finance Corporation (ILFC) ordered five 737s, five 757s, and five 737-300ERs. Other orders included four 747-400s, three 777-200IGWs, and three 757-200s for British Airways. The Polish airline LOT were contracted to take one 767-300ER and B.737-400; Airtours were to take a 737-400; Maersk Air were to take three 737-500s; and Air Berlin were to take a 737-400. Philippine Airlines also signed a deal for seven 747-400s powered by GE CF6-80C engines.

Airbus Industrie, Boeing's European competitor, announced sales of thirty-two aircraft at the show. South Korea's Asiana, a new Airbus customer, ordered eighteen A321s; FedEx ordered eleven A300-600F freighters (mostly for use on US domestic routes); and Cathay Pacific proposed to expand its four-engined A340-300 fleet to nine with a new order. Airbus displayed its 140th A330, the aircraft painted in the markings of Malaysian Airlines.

McDonnell Douglas also strengthened its order book, Brazil's VASP ordering an additional MD-11—its ninth. The Egyptian carrier Heliopolis purchased an MD-90 with options for two more, while KTHY of Turkey ordered MD-90s with options on a further three, and TWA ordered ten MD-80s.

15

Farnborough's Fiftieth

Farnborough '98, held from 7–13 September, produced the usual mixture of weather and types. It was the fiftieth anniversary of the show's permanent relocation to its Hampshire venue in 1948. The mostly abysmal September weather forced the show's dates to move to the month of July in 2000. To celebrate the fifty years of the SBAC show, Tony Blair became the first Prime Minister in a decade to open proceedings at Farnborough. In addition, an assembly of historic types spanning those years formed a static exhibition during the trade days and part of the flying display during the final weekend. Both trade and public visitors were able to view some of the first-and-second-generation airliners in the static park, such as the famous Boeing 707, two BAC 1-11s, and a Lockheed TriStar.

On the military side, a number of classic jets were shown, including the BAC (EE) Canberra B2/6 (serial WK163), which had been demonstrated at the SBAC '56 show as the Scorpion Rocket Motor test-bed. Also shown were the Hawker Hunter, DH Vampire and Venom, Gloster Meteor and BAe Jet Provost, a number of which had already become familiar sights at the numerous British air shows held each summer. Included amongst them was the Martin-Baker Gloster Meteor T7 ejection-seat test-bed and demonstrator.

In contrast, the show revealed more than twenty new types, spearheading a display of technology far removed from that which had appeared at the first Farnborough show back in 1948. A glimpse further into the future was provided by BAe's chief executive, John Weston, on the opening day of the show, when he stated:

> ...the aerospace and defence business will become global in the next ten to fifteen years. It will consist of two or three major players in the same way as the American aerospace industry is made up today. What we in British Aerospace and our partners across Europe are doing is to create an entity, which will be powerful enough to do business with the American majors on an equal footing in the future...

> When the Airbus partners submitted their report to the governments of France, Germany, Spain, and the UK in March, there was a remarkable degree of agreement on the goal we are trying to reach—an integrated European Aerospace & Defence Company, privately owned with a distributed shareholder base, and organised to retain national identity in the countries in which it operates. Such a degree of agreement would have been unthinkable only three years ago.

In total, over 1,100 companies from thirty-one countries exhibited at Farnborough '98, taking all the available exhibition space—a sign that the aerospace industry at last was climbing out of the recession experienced earlier in the decade.

Five days before the show began, as a prelude, ATC cadets from the London and South East region had the honour of officially unveiling the Eurofighter as the 'Typhoon' to the world's press at Farnborough. Eurofighter chairman Filippo Bagnato said that the name was not only a natural follow-on to the participating countries' Tornado programme that preceded it, but also that it had a simple translation into other languages across the world. However, it was intended that the name 'Typhoon' would only be used to market the aircraft to countries outside the four European airframe-manufacturing companies—the UK, Spain, Germany, and Italy, for whom the Eurofighter name would still be used, unless these operators later introduced a name of their own.

There were reports that the announcement of the name had been delayed by German objections resulting from the association with the British Hawker Typhoon of Second World War tank-busting fame. These reports led to Eurofighter's managing director stating that there had been an accord throughout the extensive and thorough name-selection process. He said that the long timescale between speculation on the name and its final form was due to the time required to establish the aircraft's new identity as an effective marketing tool on brochures, stationery, and the other myriad items required in support of export promotion. He added that 'Typhoon' was a name with a proud heritage in both Britain and Germany, belonging to both the RAF wartime fighter and the Messerschmitt Bf 108 'Taifun' (Typhoon). The latter evolved into the outstanding Bf 109 fighter, which held equal historical significance to the Hawker Typhoon. In the event, the name 'Typhoon' was adopted universally.

A new innovation at the show was the SBAC's UK 'Youth in Aerospace' theme during the public days of the show. On these days, the organisers waived the entrance fee for all children under sixteen. Robert Gardner, BAe's Vice President of Exhibitions and Promotions, provided the Corps with exhibition space around the BAe pavilions, and he wished good luck to the ATC through Cadet Warrant Officer Wendy Hill, of 457 Farnborough Squadron, in meeting BAe's 'Millennium Challenge'. In this, the company promised the ATC £100,000 if it could raise an equivalent amount by the start of the next SBAC show, in 2000. In addition, former Harrier test pilot John Farley obtained substantial BAe sponsorship for his idea of school aviation camps, where an annual flow of 150 suitably qualified students would attend a one-week course at the Cranfield Aerospace Academy in Bedfordshire. Rolls-Royce also announced its support for a University Technology Partnership designed to attract graduates into the aerospace industry.

During the show, ATR confirmed the Oman Air contract was for two ATR42-500s plus options on four more, with the whole deal valued at $90 million. The first aircraft—to be delivered in December 1998—brought the number of ATR42s sold to 343 and 343 delivered.

One of the major orders at the '98 show came from the world's biggest package-delivery company, United Parcel Services (UPS), which selected the Airbus A300 freighter to replace its Douglas DC 8s. UPS announced it would order thirty new-build A300-600Fs and take an option on thirty more; the deal would be worth $6 billion if the options were taken up. The choice of engine was between the GE CF6-80C2 and the Pratt & Whitney PW4158. Deliveries to the UPS were to begin in 2000. The order brought the number of Airbus A300 sales up to 518.

GECAS placed an order for thirty Airbus A320-200s and took options on fourteen more—all powered by CFM-56-5 engines. These firmly ordered aircraft were to be delivered between 2003 and 2006. The American leasing company ILFC also ordered six A320-200s at the show, but it did not announce which engine had been chosen. Dubai-based Emirates became the Middle Eastern launch customer for the four-engined ultra-long-range Airbus A340-500 with an order for six that was worth $2.4 billion, with further options for ten more. The airline chose Rolls-Royce Trent 553 engines to power its 500 Series, with the contract worth $640 million to Rolls-Royce if the options were exercised. ILFC announced an order at the show for ten A340s that could be -500 or -600 variants. Deliveries were to begin in 2002, with the aircraft to be powered by Rolls-Royce 60,000-lb static-thrust Trent 500 engines worth $400 million.

Boeing's 737-based Business Jet BBJ made its first flight on 4 September 1998, spending two hours and two minutes in the air. At Farnborough, Boeing announced that the aircraft would sell for around $40–45 million fully furnished. With six 737-700 BBJs sold at the show, the total orders at this time stood at thirty-five. SAS converted two options for two 737-600s into firm orders, bringing the number of aircraft bought to forty-three. Other orders for Boeing aircraft included twelve B.737-700/800s for SAS (powered by CFM56-7 engines) at a cost of $150 million and a mega-order from Varig for thirty-nine Boeings, in a deal that was worth $2.7 billion. A major element of the Varig contract was the inclusion of six Boeing B.767-300ERs. Varig's order for four Boeing 777 twin-engined wide-bodies meant that the carrier would become the first South American airline to operate the type. Elsewhere, KLM became the launch customer for the 170-seat 737-900, making an order worth $230 million for four aircraft; the company already operated thirty-six 737-300s and -400s, with eight series -800s on order.

Bombardier/Canadair announced that Atlantic Southeast Airlines had placed an order worth $575 million for fifteen Canadair RJ200s and twelve seventy-seat RJ700s, with the RJ700 order book standing at forty-one.

The Boeing Business Jet BBJ is a development of its 737-700 'Baby' Boeing and can be fitted to individual requirements. (*Author*)

British Aerospace sold three Avro RJ100s to bring total RJ sales to 150, with 111 delivered. By this time, with a total of 370 146/RJs sold to fifty-five operators worldwide, the design had become Britain's most successful civil jet project ever. The type secured a 37 per cent share of the 70–100-seat sector of the market. The turboprop side of the market had recovered, and BAe Leasing had only thirty-five of its 379 turboprop aircraft idle in 1998.

Embraer's new ERJ-135 thirty-seven-seat regional jet has 90 per cent commonality with its larger ERJ-145. At Farnborough, the company revealed that American Eagle had agreed to purchase seventy-five ERJ-135s and had taken options on seventy-five more; the deal was potentially worth $2 billion to Embraer and $500 million to Rolls-Royce for their Allison AE 3007 engines. Meanwhile, Rolls-Royce announced that Qantas would upgrade the RB 211 engines on its existing Boeing 747-400 fleet to the latest RB 211-524G/H-T standard.

On the military side, BAe and Boeing announced that they had agreed a teaming agreement to supply four Boeing C-17 Globemaster IIIs to meet the RAF's interim long-range strategic transport requirement pending delivery of the Airbus Military A400M (formerly FLA, now named 'Atlas' in RAF service) medium-range tactical transport.

On the third day of the show, Lockheed received FAA certification for its new C-130J Hercules II, with eighty-three on order—including twenty-five for the RAF, the first of which had been delivered two weeks before the show began. One was subsequently lost to a shoulder-launched SAM in southern Iraq early in the 2000s.

At the other end of the military spectrum, Boeing's F-22 air dominance fighter had completed sixty of the 183 flight test hours it was required to before the end of 1998. This was a prerequisite for the release of funding for the procurement of long-lead items for the first series production improved F/A-22A Raptor multi-role aircraft.

Marconi (formerly GEC-Marconi) had been developing the Brimstone anti-armour missile for two years, with the missile on order to the RAF. Similar in appearance to the Hellfire air-to-surface missile, Brimstone was to undergo 'drop' tests from a Tornado IDS aircraft, after which it would undergo initial firings in 1999; the target in-service date was 2001. Tests of the warhead and the seeker unit had already been carried out. Marconi was the leader of the Euroradar consortium, which sought to produce the Eurofighter's ECR-90 radar (later named 'Captor'). The radar was originally developed by Ferranti at its Edinburgh plant and air tested in a former RAE BAC 1-11. As such, Marconi displayed the performance of the system in one of their two BAC-11 trials aircraft exhibited at Farnborough; the other played a crucial role in developing the latest air traffic control system technology on behalf of the National Air Traffic Services (NATS and Eurocontrol), allowing the old air traffic control centre at Uxbridge, near London to close.

The Russian contingent was again down on previous years, but the Sukhoi Su-27 Flanker flown by Anatoly Kvochur and two MiGs made up for the absentees. The latest Mikoyan MiG-29SMT Fulcrum attack version and the two-seat multi-role UBT appeared in both the static and flying displays. The UBT was a new development of the basic UB trainer, offering the same weapons capability as the single-seater while also having a weapons system operator in the rear cockpit. As with the SMT, MiG MAPO was offering to supply new-build aircraft for the two-seater, upgraded UB trainers.

The main types on display at Farnborough '98 included:

Boeing B-52H Stratofortress (the first time ever that a B-52 bomber was displayed at Farnborough)
Boeing B-1B Lancer from USAF Air Combat Command (ACC) (one of the original Rockwell B-1A prototypes had appeared at the Farnborough '82 static display fifteen years earlier)
Boeing C-17 Globemaster III
Boeing B.2A Spirit bomber
Boeing F-15E Strike Eagle
Boeing Sikorsky RAH-66 Comanche multi-role helicopter (the type's world debut—it was cancelled in 2004)
Boeing/Aero L-159 Advanced Light Combat Aircraft ALCA (displayed for the first time outside the Czech Republic)
Dassault Rafale
Eurofighter Typhoon
Grumman F-14B Tomcat (from the US Navy, coordinated by GEC Avionics with a new digital flight-control system.
IAI Galaxy (the second aircraft, still in bare metal and primer; it was taken off flight test specifically for the show)
Lockheed Martin F-16C Fighting Falcon

Lockheed Martin F-117A Nighthawk stealth fighter
Lockheed L-1011 TriStar (equipped as a flying hospital to carry up to six intensive-care units on board)
MDC F/A-18A Hornet
Mikoyan MiG-29SMT Fulcrum
Mikoyan MiG-29UBT Fulcrum
Grumman E-2C Hawkeye (US Navy)
General Grumman Atomics Predator UAV
Saab JAS 39 Gripen
Saab JAS 39B Gripen (dual-seat variant)
Sukhoi Su-27PD '598' Flanker

This time, the Flanker was sponsored by a number of Russian companies; Sukhoi itself wanted to use its government-agency funding allocation to travel to Zhuhai and Bangalore, where shows were more likely to produce prospective customers.

Aircraft that were new to Farnborough included:

Antonov An-74TK STOL jet transport
Airbus A330-243 long-haul airliner
Bombardier Global Express (a long-range corporate jet used for Raytheon/RAF battlefield ASTOR system. The modified version, the Sentinel R1, is now operated by No. 5 Squadron, based at Waddington, Lincolnshire.
Embraer ERJ-135 corporate/regional jet
Kazan Helicopters Ansatt light multi-purpose helicopter
Lockheed Martin C-130J Hercules II
PZL-Mielec M-26 civil and military aerobatic trainer
PZL-Mielec M-28PT Skylark (a Polish version of the Antonov An-28 STOL transport)
Tupolev Tu-204-120 (which had made its debut at the Dubai show in November 1997)

Other types in the vintage category included:

Avro Avian III (shown on 13 September on the first stage of its flight to Australia)
Antonov An-2 Colt
Lockheed C-121A Constellation (USAF Material Air Transport Command 'MATS Connie')

At the 1998 show, a reliable Russian aviation industry source revealed that the reason for the ever-diminishing numbers in the Russian contingent lay with a central Russian government agency that was responsible for allocating funds to aerospace companies wishing to send aircraft to overseas international aerospace exhibitions. The agency was allegedly in the habit of only releasing money for such

The Boeing B-52 in its H-model variant. The type is destined to remain in service until 2040. (*Author*)

The Boeing B.1B Lancer is the upgrade development of the earlier cancelled Rockwell B.1A. (*Author*)

The Grumman F-14 Tomcat air-superiority fighter in its upgraded 'Bombcat' variant. It was a truly formidable weapon in the US Navy's carrier-borne inventory for almost thirty years. (*Warplanes collection*)

Despite its futuristic appearance and stealth qualities, under its skin, the Lockheed F-117 Nighthawk was basically an F/A-18; the Nighthawk used many Hornet components. (*Author*)

The multirole Boeing F/A-18E/F Hornet forms the backbone of the US Navy's carrier-borne strike force in the new millennium. (*Author*)

Predator Unmanned Aerial Vehicles (UAVs) are operated by the reinstated No. 39 Squadron (RAF) in the RAF's twenty-first century battle order. (*Aviation News*)

purposes just a few days before each exhibition opened. In reality, this led to the funding to take aircraft to overseas shows becoming virtually non-existent.

Hush-kits were much in vogue again at Farnborough '98, with many airlines needing to keep their older types in service compliant with Stage III noise requirements. The regulations particularly affected the regional and international freighters of overnight parcel carriers in the UK and the USA. Two companies producing hush-kits had aircraft on display at Farnborough '98; firstly, Burbank Aeronautical Corporation II showed their Stage III hush-kit on Boeing 707-3J6B US-registration N717QS. Their modifications for JT3D-3B and JT3D-7 engines met Stage III requirements for take-off, side-line, and approach, and the first conversion was due for completion in October 1998. Conversions take about one month and cost about $3 million for four engines. The other hush-kit at the show was displayed fitted to the starboard Rolls-Royce Spey 512-14D engine of European Aviation Ltd BAC 1-11-510ED UK civil registration G-AVMM. The hush-kit had been developed jointly by EAL and Quiet Technology Venture Inc. and featured a new acoustic lining in the nose cowl plus a twelve-lobe fluted mixer nozzle and jet exhaust noise suppressor ejector. Flight tests were completed on a BAC 1-11-400 aircraft and ground tests on the series -500, with the completion of flight tests on this aircraft made in October 1998.

British Aerospace unveiled its contender for the Lead-In Fighter Trainer (LIFT) market, a version of its successful Hawk, on 7 September, the opening day of the show. BAe were keen to emphasise that the aeroplane was very different to the previous models and not just a makeover of an old aircraft; it admitted that it even considered changing the name to reflect this. Despite this, the aircraft is the same basic airframe that began life as the Hawker Snidely Hawk 1182, the prototype of which made its first flight on 21 August 1974. However, the new Hawk benefitted from a comprehensively updated range of equipment and a 'combat' wing. It was basically similar to the Hawk Mk 127 supplied to Australia, but with many additional items.

Emphasising its commitment to the new Hawk 100, BAe invested £20 million in improved production facilities at its Brough factory. The Hawk LIFT features a full-glass cockpit, new avionics, an inertial navigation unit and GPS, an on-board oxygen-generating system, an NVG-compatible cockpit, new HOTAS, and a range of optical items including in-flight refuelling capability, FLIR, laser range-spot tracker, helmet-mounted displays, and the capability of carrying the AGM-65 Maverick missile. The two multi-function displays fitted to the RAAF aircraft are expanded into three full-colour MFDs in the new LIFT Hawk, and the Mil Std-1553B database makes the aircraft easily adaptable to incorporate future developments. At around £8–12 million per aircraft, BAe considers the LIFT Hawk good value for money. Speaking at Farnborough on 7 September 1998, BAe's chief executive, John Weston, stressed that interest in the Hawk was as high as ever, with now more than 700 sold to seventeen countries. He went on to state

The original HS Hawk T.1 first flew on 21 August 1974. Having been ordered 'off the drawing board', it was immediately adopted by the Red Arrows RAF display team. (*Aviation News*)

BAe's Hawk 100 has been selected as the RAF's LIFT for Eurofighter Typhoon student pilots. (*Aviation News*)

that he awaited the outcome of the trainer selection in South Africa; the country later confirmed the purchase of twenty-four, most of them to be built locally by Denel. Weston was confident in securing additional orders in the near future from Brunei and UAE, while additional sales campaigns were focussed on the Middle East, where discussions continued with Bahrain, Egypt, Qatar, and Saudi Arabia. Closer to home, Greece, Poland, and Slovakia were all seen as potential customers.

In terms of other military projects, Weston believed that the twenty-first century would be the age of the Eurofighter Typhoon, with Greece, Norway, South Korea, and Singapore interested. All later dropped out of the running. Greece cancelled an order for sixty, while South Korea and Singapore opted for the Boeing F-15S Strike Eagle instead. Although Australia had expressed an interest, the country opted for the Boeing F/A-18E/F Super Hornet to replace its earlier MDC F/A-18C/D Hornet variants.

All BAe Systems aerospace and defence projects were now with international partners, while 87 per cent of sales were exports; the order book stood at £22.1 billion. On BAe's link with Saab and the export Gripen, Weston stated that the company was 'pressing hard in Chile, the Philippines, and South Africa'. However, Chile later opted for the Lockheed Martin F-16. South Africa did eventually select the Tranche 3 Saab JAS 39C Gripen. All this goes to show how tenuous the market for new military aircraft is in the twenty-first century.

Farnborough International 2000 was held from Monday 24 July to Sunday 30 July. As usual, the gates opened each day at 9.30 a.m. and closed at 6 p.m. The flying displays on press and trade days were held between 2 p.m. and 5 p.m. On the public days, an extended flying programme commenced at 12.30 p.m. and concluded at 5 p.m.

The exhibition space had sold out in November 1999, eight months before the event. The SBAC provided an additional 13 per cent of indoor floor space compared to the 1998 show, bringing the total covered area to 27,634 square metres (297,460 square feet).

The show dates had been brought forward so that it ran concurrently with the UK's annual Royal International Air Tattoo, which, for one year only, was held at RAF Cottesmore, Rutland, due to runway work being carried out at its traditional home of RAF Fairford, Gloucestershire. Putting the two biggest aviation events back-to-back enabled joint organisation in many duplicated areas, with considerable collaboration between the two organising bodies. The collaborators dubbed the turn of the century events 'A Flying Start to the New Millennium'.

Amongst the new types making their debut at the Farnborough show were Boeing's 767-432ER (registration N827MH) and the BBJ. The Boeing 767 was painted in a special 'leading the way' colour scheme. It was eventually delivered to Delta Air Lines. The planned display of two Bell/Boeing MV-22 Osprey tilt-rotors did not take place because the type had been grounded for two months after a fatal crash in Arizona during operational evaluation trials.

Two Boeing F/A-18E/F Super Hornets did appear at Farnborough for the first time—one on static display and the other flying. Alongside them were the first two examples of the LMATTS C-27J Spartan transport, a re-engined and upgraded version of the Alenia G.222, which were exhibited by Alenia—one flying and one in the static park.

Various historic types were on display again in the historic aircraft park, and a number participated in the flying displays on the public days under the theme of 'Evolution of Aircraft During the Previous Century'. The flying commenced on the first public day with a replica of a 1909 Bleriot XI followed by British, French, and German representatives from the First World War. Inter-war types then followed, including a replica of Charles Lindbergh's *Spirit of St Louis* and a 'flutter' of Moths. The appearance of the de Havilland aircraft marked seventy-five years since the first flight of the popular series of light aeroplanes. The next aircraft to be shown were a selection of British and American types from the Second World War, before the display ended with a selection of early post-war piston-engined and jet aircraft.

The historic static and flying types included:

Airco DH.2
Avro Tutor
Avro Lancaster I (from the 'Battle of Britain' Memorial Flight)
Avro Vulcan (nose section only)
Bell 47G
Blackburn Monoplane

The LM/Alenia C-27J Spartan continues to secure new orders in the 2010s, particularly as a replacement for the G.222. (*Author*)

With its 'fighter-like' speed and agility, the Gates Learjet is used across the world by the USAF for communications duties. (*Author*)

The Boeing F/A-18 Super Hornet was selected by the RAAF to supplant its F/A-18C/D models for its air defence in the new millennium. (*Air Power International*)

Bleriot XI
DH Cirrus Moth
DH Sea Vixen FAW.2
DH Vampire T.11/T.55
DH Venom
English Electric Canberra
Fairey Swordfish II (FAA Historic Flight)
Fokker Eindecker E.III
Folland Gnat T.1
Gloster Gladiator
Gloster Meteor NF.11
Hawker Hind
Hawker Hurricane (from the 'Battle of Britain' Memorial Flight)
Hawker Sea Hurricane (from the Shuttleworth Collection, Old Warden, Bedfordshire)
Hawker Hunter F.6A
Hawker Hunter T.7
Hunting Jet Provost
Morane-Saulnier N
North American B-25 Mitchell
Percival Sea Prince
Republic P-47 Thunderbolt
Sopwith Camel (two replicas)
Supermarine Spitfire × 2
Westland Lysander
Westland Wasp

The main types at Farnborough International 2000 included:

Aero L-159A/B Albatross (Czechoslovakia)
Alenia AMX International A-1 (Italy/Brazil)
Alenia ATR 42MP Surveyor (France/Italy)
Alenia ATR 72-500 (France/Italy)
Alenia/LMATTS C-27J Spartan (Italy/USA)
AviaBaltika Aviation Sukhoi Su-29 (Russia)
Avione Craiova SA IAR-99 Siom (Romania)
BAe 146-100 regional airliner (United Kingdom)
BAe Systems Harrier GR.7 MDC/BAe (appearing only on Saturday 29 and Sunday 30 July, flying)
BAe Systems HS Nimrod MR.2 (trade days only)
BAe Hawk Mk 65A
BAe Systems Eurofighter Typhoon (one flypast only)

Boeing Business Jet BBJ
Boeing AH-64D Apache Longbow (USA/United Kingdom)
Boeing 767-400ER (USA)
Boeing F/A-18F Super Hornet (USA)
Boeing F/A-18E Super Hornet (USA)
Bombardier Challenger 604 (Canada)
Bombardier CRJ700 Canada (making its debut at Farnborough; the aircraft's first flight took place on 27 May 1999)
Bombardier Global Express (Canada)
Bombardier Learjet 31A/45/60 (Canada)
Bristow Helicopters Sikorsky S-61N (UK Coastguard)
Cessna 550 Citation Bravo (USA)
Cessna 560XL Citation XL (USA)
Cessna 750 Citation X (USA)
Cessna 525 Citation CJ1 (USA)
Cessna 208 Grand Caravan (USA)
Cessna 172 Skyhawk (USA)
Cessna 182S Skylane (USA)
Cessna 206 (USA)
Dassault Falcon 900EX (France)
Dassault Falcon 2000 (France)
Embraer EMB-120 Brasilia (Brazil)
Embraer ERJ-135 (Brazil)
Embraer ERJ-140 (Brazil)
Embraer ERJ-145 (Brazil)
Ericsson Saab 340 Erieye (Sweden)
Fairchild Dornier 328JET (Germany)
Galaxy Aerospace Astra SPX
Galaxy Aerospace Galaxy
GKN-Westland Lynx HMA-8 (Royal Navy)
GKN-Westland EHI Merlin HC.3 (No. 28 Squadron RAF)
GKN-Westland WAH-64D Apache AH.1 (British Army Air Corps)
Gulfstream Aerospace Gulfstream V (USA)
Gulfstream Aerospace Gulfstream IV-SP (USA)
Grob 103 Viking T.1 (Germany)
Grob 109B Vigilant T.1 (Germany)
Grob 115E Tutor (Germany)
Slingsby Firefly T67M-260 (United Kingdom)
Lockheed C-130J-30 Hercules II (USA)
Lockheed Martin F-16C Fighting Falcon (USA)
MD Helicopters MD 600N
MD Helicopters MD 902

Piaggio P.180 Avanti (Italy)
Pilatus PC-12 (Switzerland)
Raytheon/Cirrus Beech 1900D (USA)
Raytheon/Cirrus Beechjet 400A (USA)
Raytheon/Cirrus Hawker 800XP (USA)
Raytheon/Cirrus Beech King Air 200 (USA)
Raytheon/Cirrus Beech King Air 350 (USA)
Raytheon/Cirrus Beech King Air C90B (USA)
Saab JAS 39 Gripen (Sweden)
Saab/Nyge Aero Mitsubishi Mu-2B (Sweden/Japan)
Sikorsky S-76+ (USA)
Sikorsky S-92 nineteen-seat Helibus (USA)
USAF Boeing B-1B Lancer (fly-by only)
USAF Boeing C-17A Globemaster III
USAF Gulfstream C-20
USAF Boeing F-15E Strike Eagle
US Army Sikorsky UH-60L Black Hawk
US Navy Grumman E-2C Hawkeye
US Navy Lockheed P-3C Orion

Above: The Lockheed P-3C Update III Orion serves the USA and other air forces in a number of roles in addition to maritime patrol and ASW. (*Air Power International*)

Below: The Grumman E-8 JSTARS (based on former airline Boeing 707 airframes) was offered to NATO and the UK, but the latter rejected the aircraft as it was too big for the UK's needs. (*Grumman show hand-out*)

16

The New Millennium: Farnborough's Sixtieth Birthday

By the time of Farnborough international 2002, the aerospace industry had changed considerably in the aftermath of 9/11. With the airline industry in disarray and many older airliners laid up or withdrawn from use due to the downturn in air travel, the show focussed on the military and their unmanned aerial vehicles (UAVs).

Aerosystems International (Ael) joined Lockheed Martin UK Integrated Systems' team in bidding for the UK Watchkeeper UAV programme. Ael was a 50:50 joint venture between BAe Systems and GKN Aerospace, which proposed to supply systems engineering and integration for the software facility for integration with Watchkeeper.

Watchkeeper was to provide intelligence, surveillance, target acquisition, and reconnaissance at battle group, brigade, and divisional level for the British Army. The system would consist of UAVs and a system for distributing and exploiting the intelligence information. The contract for Watchkeeper was estimated to be worth $775 million.

Lockheed Martin also confirmed it had selected the EADS Eagle to meet its long-endurance UAV requirement and that it was investing company funds in evolving its sea-based UAV to meet a US Navy requirement for an unmanned combat air vehicle (UCAV).

With a proposed in service date of 2008–09, the US Navy initially ordered two Northrop Grumman RQ-4A Global Hawks in 2003 for operational evaluation. The aircraft was to be equipped with different sensors including 360-degree radar, electronic support measures for target cueing, and a maritime-mode moving-target-indicator radar. The US Navy was considering a range of alternative UCAVs at the same time.

Saab unveiled an UCAV—one of a series the company was developing as part of Sweden's network-centric defence programme. It was revealed that Saab had built the UCAV to Swedish military aircraft standards rather than as a model aircraft to develop experience in building operational UCAVs. One of Saab's aims was

In the new millennium, the Grumman RQ-4 Global Hawk has been evaluated by the USAF. (*Air Power International*)

to develop UAVs with systems and levels of airworthiness that allowed them to operate routinely in civilian-controlled airspace.

The tailless design was shown at Farnborough 2002 with the engine on the upper rear fuselage. Although the demonstrator displayed was manually controlled, it was used to prove flight-control systems for UAVs (including autonomous take-off and landings) and UAV sensors.

As well as UCAVs, Saab has developed proposals for a high-altitude long-endurance UAV equipped with Ericsson Erieye radar and unmanned reconnaissance air vehicles that could be launched from the company's JAS 39 Gripen to provide pre- and post-strike imagery. Saab is also working on command and control systems to provide decision makers with an integrated picture of air, land, and sea sensors.

BAe Systems Avionics—formerly GEC/Marconi Avionics—displayed its first airborne active electronically scanned array (AESA) radar for maritime helicopters, UAVS, and the Seaspray 7000E. The radar is also suitable for small maritime-patrol aircraft. The AESA-equipped multi-mode radar has a 'much smaller' mass and volume and simpler installation than equivalent mechanically scanned radar.

Meanwhile, BAe Systems displayed its mock-up Eurofighter Typhoon, which was fitted with conformal fuel tanks for the first time. Earlier in 2002, BAe concluded high-speed wind tunnel tests of the tanks, each increasing the type's fuel capacity by 1,500 litres (400 US gallons) and together extending the aircraft's range by around 350 km.

On the civil side, Bombardier revealed it had started work on its Learjet 40 light business jet—a shortened version of its super-light Learjet 45. Meanwhile,

A selection of European UAVs were displayed. (*Author*)

the company was also upgrading its Learjet 45 to have a higher take-off weight to increase range and installed uprated engines to improve hot-and-high performance. The maximum take-off weight was increased by 450 kg (1,000 lb), which allowed the re-designated 45XR to carry eighty passengers and a full fuel load to give an 830-kn (450-nm) range advantage over the rival Cessna Citation Excel.

Virgin Atlantic was the launch customer for the four-engined Airbus A340-600, and the company presented its first aircraft at the show ahead of its planned service entry on the LHR-New York JFK route. The 311-seat airliner was the first of ten of the Rolls-Royce Trent 556-powered aircraft that Virgin was scheduled to take delivery of through to June 2006.

On 23 July at the Farnborough 2002 show, FedEx confirmed that it had selected the GE/Pratt & Whitney Engine Alliance GP7277 to power its planned new fleet of Airbus A380-800 freighters. The engine deal had been confirmed in the previous week, when FedEx had firmed up its memorandum of understanding for ten A380-

800Fs, plus options on a further ten. The first three aircraft were scheduled for delivery in 2008, followed by three more in both 2009 and 2010 and the final aircraft in 2011. However, the order was subsequently cancelled due to delivery delays.

In 2002, Airbus had ninety-five orders in hand from eight carriers and another fifty-seven options for its new freighter version of the A380 'Super Jumbo'. At this time, Rolls-Royce had been selected to power forty-three aeroplanes. Airbus's Trent 900-powered aircraft were ordered by Lufthansa, Qantas Airways, Singapore Airlines, and Virgin Atlantic. The Engine Alliance GP7200-powered aircraft had been ordered by Air France, Emirates, and FedEx.

At Farnborough International 2004, Thales announced that it had secured the order to deliver the Watchkeeper UAV intelligence, surveillance, target acquisition, and reconnaissance system. The deal was worth around $1.5 billion (£800 million) and included integration of the Thales I-Master light-weight synthetic-aperture radar/moving-target-indicator payload into the WK180 low-level UAV. Thales aimed to make its first deliveries of the aircraft by late 2006.

Behind the flurry of announcements for civil types, Airbus and Boeing executives exchanged verbal 'brickbats' on the subject of large-aircraft subsidies in the United States and Europe, although much of the talk in the chalets and halls was about the proposed future of the show itself and the proposed revamp for exhibitors.

With regards to Boeing's new 787 Dreamliner, it was announced that there would be a design freeze in July 2005, with the first 787 scheduled to fly in the third quarter of 2007 and enter service a year later. To hasten series production, Boeing issued contracts for three Boeing 747-400 Jumbo Jets to be converted and transfer sub-assemblies to the final assembly line at Everett; the first aircraft were to be available for work in late 2006. By Farnborough International 2004, four airlines had committed to sixty-two 787s, including Air New Zealand, All Nippon Airways (ANA), First Choice Airways, and Blue Panorama. Boeing stated that it had taken deposits from approximately twenty-four airlines for over 200 aircraft.

Embraer's ERJ 190 highlighted the company's continued influx into the large regional jet market. This trend was reinforced by Bombardier's unveiling of plans for a 100 to 130-seat aircraft. Bombardier's Global 5000 was shown for the first time with a production interior.

On the military side, the show was stolen by the unveiling of full-size mock-ups of the X-45C and X-47B UCAVs. There was also speculation about a break in Eurofighter production should the Tranche 2 production order not be signed. The UK government was reportedly in talks with the American government on possible participation in the Boeing UCAV project. Boeing began assembly of the X-45C displayed at the show in 2004, with a first flight planned for mid-2006. Northrop Grumman's larger X-47B was scheduled to fly in October 2006. Each company planned to build three UCAS (Unmanned Combat Air System) demonstrators for a joint USAF/US Navy operational assessment to run from 2007 to 2009.

Boeing offered the X-45B N-UCAS (naval unmanned combat air strike) aircraft for carrier-borne operations to enter service around 2018.

Boeing's dual-seat F/A-18F Super Hornet put on a superb display and wowed the crowds. The Italian AMI received its first Alenia-modernised Tornado IDS aircraft in early July and the aircraft was immediately deployed to the UK for the Farnborough Show. The aircraft was the first of eighteen to be completed under a $44-million contract awarded to Alenia in July 2002 by the NATO Eurofighter and Tornado Management Agency through the tri-national consortium.

Lockheed Martin revealed at the show that it was to continue with its plans to repackage its P-3C Orion even though they had lost out to Boeing's 737-800ERX in their bid for the US Navy's Multi-mission Maritime Aircraft (MMA). The Lockheed Orion 21 is a 40 per cent more powerful version of the P-3C, and a new package of options will be offered to existing and new international customers, enabling them to select a degree of individuality or accept the aircraft as a standard package.

The Beech Raytheon T-6B single-engine turboprop advanced trainer made its debut at the show. Although the aircraft shared 85 per cent commonality with the Beech T-6A that won the competition for the Joint Primary Aircraft Training System (JPATS), the upgraded version added the CMC Electronics FV-4000 cockpit, which included a Sparrowhawk head-up display, multifunction displays, and a stores-management system. The T-6B introduced the capability to carry wing-mounted weapons and stores. The maximum load for the light attack version could include a mix of external weapons, tanks, and rocket pods or Mk 82 225-kg (500-lb) unguided bombs, with each wing able to carry up to 680 kg of weapons and fuel. The trainer is expected to be offered for a number of export deals, starting with a planned order by the Columbian military to replace their ageing Cessna A-37B Dragonfly armed trainers and multi-purpose COIN Rockwell OV-10A Broncos. Following its appearance at Farnborough 2004, Raytheon announced it was sending the T-6B Texan trainer on a world tour of at least eleven nations.

Sweden's SaabTech secured its first order for electronic warfare systems on the Boeing F/A-18C/D Hornet, with the RAAF signing an initial contract (worth $2 million) for evaluation countermeasures systems for the aircraft.

On the civil side, Gulf carrier Emirates concluded a deal with Boeing for up to thirteen 777-300ERs—four firm orders and nine options. The deal brought the carrier's firm commitments for Boeing 777 GE90-115B-powered aeroplanes to thirty. Etihad concluded a major fleet-expansion deal with an order for twenty-five Airbus wide-bodies. The new Abu Dhabi airline, which had launched the previous year with leased Airbus A330-200s, ordered four A340-500s, four A340-600 four-engined aircraft, and twelve A330-200 twins. The eight A340s were to be powered by Rolls-Royce Trent 500 turbofans. Two additional A330s to be leased in 2005 would also be Rolls-Royce-powered, being fitted with Rolls-Royce Trent 700s.

The Raytheon T-6 Texan that won the US JPATS competition is a highly-modified Pilatus PC.9 derivative. (*Author*)

The Airbus A318 twin turbofan short-haul airliner demonstrated its steep-approach capability during its flying display at the 2004 show. Airbus targeted certification of a glide-path of 5.5 degrees compared to the normal 3 degrees. Minor software changes had been made to the type's fly-by-wire system to achieve the required approach characteristics.

India's start-up low-cost carrier Kingfisher Airlines confirmed it would start operations between the country's major cities in early 2005 with four second-hand Airbus A320s.

Boeing announced it had reached the 95 per cent firm design milestone on its 777-200LR, which was due to enter final assembly in late 2004 and start flight testing in 2005. Following the launch of the 787 Dreamliner, the Boeing 777-200LR appeared in the new Boeing house colours. The 777-200LR would enter service with launch customer PIA in January 2006.

In Asia and the Pacific Rim, the increased demand due to low-cost carriers led Boeing to revise its projected long-term worldwide airliner demand in 2004:

Regional Jets: 4,285
Single-aisle types: 14,715
Twin-aisle types with 200 to 300 seats: 2,620
Twin-aisle types with 300 to 400 seats: 2,115

The Boeing 777 is now well-established in a number of versions across the world's airlines. (*Air Power International*)

Freighters: 724
Boeing 747 Jumbo Jets and larger: 535—only around 50 per cent in the Airbus A380 Super Jumbo category, around 270 aircraft.

Meanwhile, an independent analyst estimated that Airbus would need to sell 400 'Super Jumbos' to break even. In fact, it would take Airbus fifteen years to recover from the EADS Airbus A380 debacle.

Sukhoi Civil Aircraft was expected to receive an order from Russia's largest domestic carrier, Sibir, for up to 100 RRJ-95B short-haul regional jets. The initial deal was for up to fifty, with ten additional options; it was worth around $1 billion. The Sukhoi sixty to 100-seat RRJ family was to be offered in three fuselage lengths—the RRJ-60, RRJ-75, and RRJ-95—in basic and long-range variants, with the latter incorporating additional fuel capacity. The aircraft looked suspiciously similar to the BAe Systems/Turkish Aerospace defunct twin-engined RJ design proposal. All the aircraft are powered by Snecma/NPO Saturn SM 146 engines.

In July 2004, as part of a revamp of the Society of British Aerospace Companies, it was announced there would be a radical shake-up for the Farnborough Air

Show, with fewer trade days and a slimmed-down flying display and even possibly a new location.

The SBAC, which had run Farnborough since its inaugural show in 1948, revealed that this was in response to pressure that had particularly come from USA exhibitors, who were unhappy about the number and frequency of international air shows, and the growing popularity of niche events. There was also growing concern about the impact of the biennial show on the local community and complaints about the transport difficulties when attempting to reach the venue, which was 30 miles (50 km) from central London. The SBAC was also talking to exhibition companies about taking over the running of the event. President Colin Green said, 'We have to ask whether the SBAC is best placed to manage a show like Farnborough.'

The changes—still to be finalised—were originally due to be implemented from the 2006 event onwards; that year, however, the event would definitely be held at the famous Hampshire airfield. Colin Green stated, 'There will be a Farnborough '06, but it won't be a mirror image of '04. After that, nothing has been ruled out, including a change of site or even frequency.' However, Green admitted that finding another airfield site could be hard, given the need for access to an international airport and hotels. One possibility is a merger with the Royal International Air Tattoo—a charity military air show held at RAF Fairford (a USAF Air Combat Command forward operating base) in Gloucestershire. However, Green said that this would present problems because of the 'very different character' of the events.

In 2004, the still-biennial Farnborough was the fourth major international air show in eight months, with the Dubai air show taking place in December of the previous year, Singapore's Asian Aerospace show in February, and Berlin's ILA in May. The event was also sandwiched between two business aviation conventions—May's EBACE in Geneva and October's NBAA in Las Vegas.

SBAC Director General Sally Howes, who took over in 2003, stated that she was spearheading a society reorganisation that would 'see membership criteria widened, a new role for regional consorts representing smaller aerospace companies, and a move to a more "project-based" approach to provide member services'.

At Farnborough '06, Sukhoi renamed the Russian Regional Jet the 'Superjet 100' as part of a company-wide rebranding exercise. Meanwhile, the company selected Thales to supply support and training equipment for the type and to supply and integrate the aircraft's avionic suite.

BAe Systems Regional Aircraft announced new leasing deals. Italian regional carrier Club Air Sixgo doubled its BAe 146 fleet with the lease of two 146-200s. Columbian carrier Aerolinea de Antioquia ADA bought three Jetstream 32EPs, the aircraft replacing its de Havilland DHC-6 Twin Otters. In addition, BAe Regional secured $7 million worth of modification business in the first half of 2004, the work covering both mandatory and upgrade modifications for BAe 146/Avro RJs from customers including Eurowings, FlyBe, SN Brussels, and Swiss International Air Lines.

Ukrainian manufacturer KhGAAP won orders for ten of its twin-turbofan short-haul medium transport An-74TKs; the aircraft had been demonstrated for the first time at the Paris Air Show in June 2001. The bulk of the orders were from Egypt and Libya, with the rest from 'rich individuals'. KhGAAP stated that Egypt intended to use its An-74s for 'logistic missions', while Libya ordered the An-74TK300VIP variants primarily as VIP transports, but they would also be used as medical evacuation aircraft.

Yokohama-based ShinMaywa Industries builds and supports the world's only military operational flying boat—the only aircraft the company has ever manufactured. The type is operated by the Japanese Maritime Self-Defence Force (JMSDF), who placed an order for four upgraded US-1A search-and-rescue amphibians the company had been developing for many years. ShinMaywa was promoting its civil variant at Farnborough '04, and the company revealed that it was to start a sales drive to attract customers in the Asia-Pacific region with a requirement to combat bush fires.

AgustaWestland unveiled a stretched derivative of the A109 light helicopter at the show; it was named the 'Grand'. The version was designed to capture a share of the growing intermediate market. The eight-seat helicopter would compete against the Eurocopter EC145 and be positioned in the Agusta range between the A109 Power and the Bell Agusta AB139. The Grand was scheduled for certification as the 'A109S' at the end of 2004, with the Italian manufacturer having four aircraft on the production line at this time.

According to the SBAC, in 2004, UK aerospace sales grew 3.5 per cent in real terms (to $32 million) and employment to by 4 per cent (to just below 122,000) compared to 2003. The SBAC pointed out that this was a mild recovery after two years of depressed demand. Research and development spending increased by 18 per cent to £2.1 billion, the third rise in three years. Companies invested 12.3 per cent of revenues into research and development, compared to 8.2 per cent in 2001.

For the first time since the mid-1990s, the defence sector had a bigger share of the UK turnover than civil aerospace.

At Farnborough International '06, the forty-fifth biennial exhibition, the flying display was to be held from 17–23 July, and the exhibition would be open to the public on Saturday 22nd and Sunday 23rd. The air conditioning systems in the exhibition halls had to battle against the soaring temperatures, and the audiences in the sweltering heat outside were treated to a spectacular air display. Airbus's new chief executive, Christian Streiff—who resigned just three months later under pressure from parent company EADS's senior executive—had his first grilling with regard to the company's problems with the A380 'Super Jumbo'. It was one of the most eagerly anticipated press conferences in recent Farnborough history. Prime Minister Tony Blair also put in an unannounced appearance at the show.

The hot news included Airbus's A350 Xtra Wide-Bodied—the XWB twinjet with its ovoid (as opposed to circular) fuselage cross-section. The revamped A350 now

Four Argentine Agusta A 109s were captured by British forces during the 1982 Falklands War and returned to the UK for use by Royal Marine Commandos. (*Take off*)

The Agusta A 109 Grand is an upgraded development of the Agusta A109 Power. (*Author*)

bore more than a passing resemblance to Boeing's 787 Dreamliner. In addition, a previously secret UAV from Lockheed Martin was displayed.

A world record was established by a Sino Swearingen SJ30 light business jet flying from its base in San Antonio, Texas, to Farnborough in ten hours and twenty-four minutes, including a forty-two-minute refuelling stop at Goose Bay, Canada. The jet flew direct from Goose Bay to Farnborough, setting a speed and range record. The SJ30 received FAA certification in October 2005. First deliveries were due in 2006.

On the second day, BAe Systems confirmed that it had been given the go-ahead to produce a further nine Nimrod MRA/4A maritime-reconnaissance and attack aircraft for the RAF, The company was also told to bring the three development aircraft to the same standard; by the time of the show, those aircraft had made 320 flights between them (including night landings), operating with a two-crew flight deck and attaining an operating altitude of 10,700 metres (35,000 feet).

Raytheon displayed its new Sentinel R1 (serial XJ692) airborne stand-off radar system (ASTOR) in the static park, its first major public appearance. The first two of five heavily-modified Bombardier Global Express business jets were due to enter service with No. 5 Squadron at RAF Waddington, Lincolnshire, in November 2006. ASTOR is the UK's equivalent of the USAF's Grumman E-8 JSTARS, which is a modified Boeing B.707 airframe that was declared too large for purpose by RAF chiefs.

Lockheed Martin's 'Skunk Works' revealed its all-composite flying-wing UAV, the P-175 Polecat, which was developed and built in secret in just eighteen months and flown for the first time in February 2006. The Polecat has a wingspan of 27.4 metres (90 feet), weighs 4,080 kg (9,000 lbs) with a 455 kg payload, and is powered by two Williams FJ44-3E turbofans, producing a combined 26.7 kN of static thrust. Lockheed believes that its Polecat could lead to a subsonic unmanned strike aircraft.

Emirates explained its decision to buy ten Boeing 747-8 Freighters just months after dropping its order for the Airbus A380 'Super Jumbo' Freighter; the company stated that the Boeing aircraft was more clearly defined and better-suited to its needs. The Dubai carrier switched the order for two A380 Freighters to the passenger model instead. Emirates claimed it had not received a specification from Airbus on their freighter.

Despite the company's problems, Airbus won the day at the show, with orders for ninety-three of the A320 family and twenty-one A330s amounting to $9.7 billion. On the other hand, Boeing secured seventy orders, including requests for fifty 737s, two 747-400s, ten 747-8 Freighters, two 777 and 777F Freighters, and four 787 Dreamliners; these deals were worth around $9.1 billion in total.

In 2007, EADS claimed that every effort was being made to resolve their A380 production problems and that several customers had placed orders for 'customised' A380 variants in various luxurious, high-spec fits. Optimistically, the company stated that in due course, with an ever-expanding market in this field, the projected company sales of the aircraft would be realised.

The thrill of the show's air display was restored by the all-aspect thrust-vectoring RSK MiG-29M OVT fighter. While its double backflips and controlled spins might be regarded as display gimmicks, they are designed to show the control power provided by the three-dimensional vectoring nozzles on the Russian fighter's twin Klimov RD-33 OVT engines. Gripen International displayed a two-seat Saab JAS 39B Gripen fighter in the livery of the UK's Empire Test Pilots' School, which is using the type under a 'power-by-the-hour' deal.

Following the delivery of the first of six of its advanced BAe Hawk 129 jet trainers (AJT) for Bahrain, BAe Systems exhibited the first of its two Hawk 128 AJT demonstrators for the UK's RAF in anticipation of a twenty-eight-aircraft order. The type was to be used as a LIFT for Eurofighter Typhoon pilots. The type's sensor simulation capabilities were demonstrated using Hawk New Development Aircraft (HNDA) serial ZJ951.

AgustaWestland revealed details of its plans for the AW149 military medium twin helicopter, which it planned to fly for the first time in 2009. The 7–8-tonne helicopter was to achieve initial operating capability by 2011. The company intended to display a full-scale mock-up at the 2007 Paris Air Salon. With several thousand similar-sized military helicopters of its class needing replacement by the end of the decade, AgustaWestland looked to provide their replacements.

The Bell/Boeing V-22 Osprey tilt-rotor made a pass over the UK Houses of Parliament during the show, and the company took the opportunity to launch an international sales drive, with Australia, Israel, Japan, and the UK seen as potential customers.

Not to be forgotten, Lockheed Martin was working to define the configuration of a 'next-generation' F-16NG to meet the requirements of emerging customers, possibly including India. The final definition of this configuration was awaiting input from potential customers of their requirements, but it was intended to combine the best of Block 50 and Block 60 designs and additional elements from the F/A-22A Raptor and F-35 Joint Strike fighter. With the Indian Air Force a possible customer for up to 160 F-16NGs, the company would not decide the final configuration until India had issued its request for proposals and the US government had decided what new technologies could be released. The aircraft is particularly attractive to existing F-16 Fighting Falcon/Viper operators who do not intend to procure the LM F-35 Lightning II.

A major revelation at the 2006 Show was made by Russia's state arms vendor, which stated that the first deliveries of six Sukhoi Su-25T or Su-39 Frogfoot attack aircraft had been made to Iran. The Su-25T differs from the standard Su-25 in having the Krasnogorsk Optical Plant Zarevo electro-optical sighting system. The Zarevo enables the use of laser-beam-riding Vikhr supersonic anti-tank missiles. It is believed the aircraft were delivered with guided missiles of an unknown type.

The Lockheed-Martin f-16NG can be fitted with fuselage saddle-tanks, extending its range and endurance. (*Author*)

The deal with Rosoboronexport also included spares and upgrade kits for ex-Iraqi Air Force Su-25 Frogfoot aircraft that were ferried to Iran by Iraqi pilots during the evacuation to Iran of IrAF combat planes during Operation Desert Storm in 1991. None of the aircraft were ever returned, except for a few Boeing airliners.

The International Space Pavilion was host to a number of organisations, including the European Space Agency (ESA) and British National Space Centre (BNSC), along with a number of space-equipment-manufacturing companies. The International Space Pavilion was interactive during the air show weekend, with a series of seminars taking place that were mainly aimed at getting children involved with and excited by aerospace.

The UAV Pavilion, comprising a comprehensive display of UAVs, was new to Farnborough International Air Show in 2006. The Pavilion appeared to hold high interest and fascination with public visitors due to the UAVs' expanding use and applications, and for the obvious reason—they are unmanned.

In 2008, which marked Farnborough International Air Show's sixtieth birthday, the show continued to thrill and entertain all age groups, just as it had done at its inaugural Farnborough show in 1948. It was the perfect day out—not only for the dedicated aviation enthusiast, but for families as well. In the new millennium, apart from the main business of securing new orders for the 1,360 or more exhibiting companies that make up the global aerospace industry, there are many aviation-related attractions, such as simulator experiences and helicopter rides.

A range of delights have also been introduced to keep the young ones interested, such as extreme rides, big bouncy castles, carousels, face painters, children's costume characters, crowd entertainers, bungee trampolines, a climbing wall, and much more. Other activities for children include a dedicated arts and crafts area, where they can paint, glue, and stick while the 'grown-ups' enjoy the aviation delights (both static and flying) that the show always provides.

Bibliography

Books

Burbridge W. F., *Balloon to Bomber* (John Crowther Publications, England 1946)
Endres. G., and Gething M., *Jane's Aircraft Recognition Guide* (HarperCollins, London 2002)
Hoare, R. J., *Black's Junior Reference Books: The Story of Aircraft* (A & C Black, London 1983)
Middleton, D., *Civil Aviation: A Design History* (Ian Allan Publishing, London 1986)
Simpkins, P., *Air Fighting 1914-1918* (IWM, London 1978)
Taylor, M. J. H., *The Aerospace Chronology* (Tri-Service Press, London 1989); (ed.), *Jane's Aviation Review* (Janes Information Group, London 1987)
Thetford, O., *Aircraft of the Royal Air Force Since 1918* (Putnam, London 1988); *British Naval Aircraft Since 1912* (Putnam, London 1982)

Magazines and Periodicals

Air Pictorial Journal of the Air League (Magazines and Publications Ltd, various issues)
Flight Aircraft, Spacecraft, Missiles: Farnborough Review September 1960
Flight Aircraft, Spacecraft, Missiles: Farnborough Review September 1961
Flight International VTOL Aircraft Survey: Farnborough Review June 1962
Flight International: Farnborough Review September 1962
Flight International: Farnborough Review September 1966
Flight: The Aircraft Engineer and Airships (1933)
Wings (Orbis Publishing Ltd, various issues)

Special Publications

Aviation Week Show News (souvenir, various)
Farnborough '66 (SBAC publication)
Britain in Aerospace (SBAC publication)
Farnborough (SBAC show souvenir, various)
Revue Aérospatiale (September 1994)